Gestural Communication in Nonhuman and Human Primates

Benjamins Current Topics

Special issues of established journals tend to circulate within the orbit of the subscribers of those journals. For the Benjamins Current Topics series a number of special issues have been selected containing salient topics of research with the aim to widen the readership and to give this interesting material an additional lease of life in book format.

Volume 10

Gestural Communication in Nonhuman and Human Primates
Edited by Katja Liebal, Cornelia Müller and Simone Pika

These materials were previously published in *Gesture* 5:1/2 (2005)

Gestural Communication in Nonhuman and Human Primates

Edited by

Katja Liebal
University of Portsmouth

Cornelia Müller
European-University Viadrina

Simone Pika
University of Manchester

John Benjamins Publishing Company
Amsterdam / Philadelphia

The paper used in this publication meets the minimum requirements of American National Standard for Information Sciences – Permanence of Paper for Printed Library Materials, ANSI Z39.48-1984.

Library of Congress Cataloging-in-Publication Data

Gestural communication in nonhuman and human primates / edited by Katja Liebal, Cornelia Müller and Simone Pika.
p. cm. (Benjamins Current Topics, ISSN 1874-0081 ; v. 10)
Originally published in Gesture 5:1/2 (2005).
Includes bibliographical references and index.
1. Gesture. 2. Animal communication. 3. Primates. I. Liebal, Katja. II. Müller, Cornelia. III. Pika, Simone.
P117.G4684 2007
808.5--dc22 2007020958
ISBN 978 90 272 2240 4 (Hb; alk. paper)

John Benjamins Publishing Co. · P.O. Box 36224 · 1020 ME Amsterdam · The Netherlands
John Benjamins North America · P.O. Box 27519 · Philadelphia PA 19118-0519 · USA

Luigia Camaioni (1947–2004)

We would like to dedicate this volume to Luigia Camaioni who passed away only few months after presenting her work on declarative and imperative pointing in infants at the workshop on Gestural Communication in Nonhuman and Human Primates from which this volume emerged.

We are inconsolable about losing a colleague devoted to science, with a broad interest in developmental psychology and an expertise in the nature of intentional communication in human infants. Her pioneering work contributed significantly to the understanding of communication in preverbal children and their use of gestures during early language acquisition. Her sudden, unexpected death leaves us missing her as a colleague and as a friend.

Table of contents

About the Authors

Shanley Allen, Ph.D., is an Associate Professor in the Program in Applied Linguistics and the School of Education at Boston University. Her research explores the first language acquisition of morphology and syntax, with a focus on comparing acquisition patterns across languages, as well as with children learning Inuktitut (Eskimo) in northern Canada. Her other interests include bilingual acquisition, specific language impairment, and the acquisition of co-speech gesture.

Michael Anthony Arbib was born in England, grew up in Australia and received his Ph.D. in Mathematics from MIT. After five years at Stanford, he became chairman of Computer and Information Science at the University of Massachusetts, Amherst in 1970. He moved to the University of Southern California in 1986, where he is Professor of Computer Science, Neuroscience, Biomedical Engineering, Electrical Engineering, and Psychology. The author or editor of 38 books, Arbib recently edited *From Action to Language via the Mirror System*. His current research focuses on brain mechanisms of visuomotor behavior, on neuroinformatics, and on the evolution of language.

Joanna Blake is a Professor Emeritus of Psychology at York University.

Amanda Brown, Ph.D. in Applied Linguistics from Boston University and the Max Planck Institute for Psycholinguistics, Nijmegen. Currently she is an Assistant Professor of Linguistics at Syracuse University. Her research investigates bilateral interactions between established and emerging language systems using analyses of speech and co-speech gestures in Japanese speakers of English.

Josep Call, Ph.D. in Psychology in 1997 from Emory University, Atlanta. Worked at the Yerkes Primate Center from 1991 to 1997. From 1997 to 1999 was a lecturer at the University of Liverpool. Since 1999 he is a research scientist at the Max Planck Institute for Evolutionary Anthropology and director of the Wolfgang Köhler Primate Research Center in Leipzig. His research interests focus on comparative cognition in the social and physical domains. He has published numerous research articles on primate social behavior and comparative cognition and a book Primate Cognition (w/M. Tomasello, Oxford University Press, 1997).

Olga Capirci, researcher of the Italian National Research Council (CNR), currently coordinates the "Gesture and Language" Laboratory at the CNR Institute of Cognitive Sciences and Technologies. Her research focuses on gesture and communication in typical and atypical development, neuropsychological developmental profiles and sign language teaching.

Micaela Capobianco is currently a post-doctoral fellow at the Università di Roma I "La Sapienza", Department of Psychology of Developmental Processes and Socialization. Her research focuses on the role of gestures in early language learning in typically developing children, and in atypical conditions (pre-term children), and on the use of different language assessment methodologies in clinical practice.

Maria Cristina Caselli, senior researcher of the Italian National Research Council (CNR), currently coordinates the "Language Development and Disorders" Laboratory at the CNR Institute of Cognitive Sciences and Technologies. Her research focuses on communication and language in typical and atypical development, neuropsychological developmental profiles, language assessment, and early identification of children at risk for language development.

Valerie J. Chalcraft received her M.A. and Ph.D. in Experimental Psychology from the University of Nevada, Reno. She is currently consulting in the field of applied companion animal behavior.

Annarita Contaldo, Infant Neuropsychiatrist at the ASL of Trento, Italy. She collaborated with CNR Institute of Cognitive Sciences and Technologies of Rome and with IRCCS "Stella Maris" of Pisa on research on language acquisition in typically and atypically developing children.

Reyhan Furman, M.A. is a doctoral student at the Linguistics Department, Bogazici University, Istanbul. Her research focuses on the event structure representations of monolingual and bilingual children and adults, in language and co-speech gestures. She is also interested in children's acquisition of verb argument structure and the acquisition of complex constructions.

R. Allen Gardner received his Ph.D. in Psychology from Northwestern University, with the distinguished experimental psychologist, Benton J. Underwood. Together with Beatrix T. Gardner (D. Phil. in Biology, Oxford University, with the nobelist and founder of ethology, Niko Tinbergen) he founded sign language studies of cross-fostered chimpanzees beginning with chimpanzee Washoe.

Susan Goldin-Meadow is the Beardsley Ruml Distinguished Serve Professor in the Department of Psychology and Department of Comparative Human Development at the University of Chicago. A member of the American Academy of Arts and Sciences, she has served as President of the Cognitive Development Society and is currently serving as the editor of the new journal sponsored by the Society for Language Development, *Language Learning and Development*. Her research interests are bifold: Language development and creation (the deaf children's capacity for inventing gesture systems which are structured in language-like ways) and gestures' role in communicating, thinking, and learning (with a special focus on gestures conveying information that differs from the information conveyed in speech). She has recently published two books representing these two venues of research: *The resilience of language: What gesture creation in deaf children can tell us about how all children learn language,* Psychology Press, 2003; and *Hearing gesture: How our hands help us think,* Harvard University Press, 2003.

Juan-Carlos Gómez is Reader in Psychology in the University of St. Andrews, United Kingdom. He graduated and obtained his Ph.D. in psychology at the Universidad Autónoma de Madrid, Spain, in 1992. In 1995, he was visiting scientist at the MRC Cognitive Developmet Unit, London. In 1996 he moved to the University of St. Andrews, where he teaches Developmental Psychology. He is member of the Center for Social Learning and Cognitive Evolution, and conducts research on intentional communication in human and non-human primates. He is the author of *Apes, monkeys, children, and the growth of mind*, Harvard University Press, 2004.

William D. Hopkins, Ph.D. (Psychology) from Georgia State University in 1990. Research Associate in the Division of Psychobiology, Yerkes Primate Center, since 1989. Research Associate at the Georgia State University Language Research Center since 1994. Associate Professor of Psychology at Berry College, Rome, Georgia, from 1994–2006. Associate Professor of Psychology at Agnes Scott College, Decatur, Georgia, since 2006.

Sotaro Kita, Ph.D., is Reader in the School of Psychology at the University of Birmingham. His main research interests are cognitive psychological, interactional, and ethnographic studies of the relationship between speech and spontaneous co-speech gestures. In addition his research interests include child language acquisition, semantics and pragmatics of spatial expressions, and cross-linguistic studies of spatial conceptualization.

David A. Leavens, Ph.D. (Psychology) from the University of Georgia in 2001. Since 2000, Lecturer in Psychology and Director of the Infant Study Unit at the University of Sussex.

Katja Liebal, Ph.D. (Biology) from the University of Leipzig and Max Planck Institute for Evolutionary Anthropology, Leipzig. Currently she is a lecturer at the University of Portsmouth. Her interest is in social communication and socio-cognitive skills in gibbons and great apes.

Ulf Liszkowski Ph.D. (Psychology) from the University of Leipzig. He conducted his doctoral and post-doctoral research at the Max Planck Institute for Evolutionary Anthropology, Leipzig, and is currently leader of an independent junior research group hosted at the Max Planck Institute for Psycholinguistics in Nijmegen. His current interest is in the ontogeny of human communication, social cognition and cooperation with a focus on prelinguistic development.

Dario Maestripieri earned his Ph.D in Psychobiology from the University of Rome in 1992 and is currently an Associate Professor at the University of Chicago. His research interests focus on the biology of behavior from a comparative perspective. He is the author of over 100 scientific articles and editor of the book *Primate Psychology* (2003).

Cornelia Müller holds an M.A. in General, German, Spanish, and French Linguistics, a Ph.D. in Linguistics and Psychology, and a Habilitation in General and German Philology. She is a Professor for Applied Linguistics at the European-University at Frankfurt (Oder). She published several articles and a book on co-verbal gestures, their semiotic structures, their cultural history, theory, and their cross-cultural comparison: *Redebegleitende Gesten: Kulturgeschichte, Theorie, Sprachvergleich* (Berlin Verlag Arno Spitz, 1998) and prepares another volume for publication: *Metaphors. Dead and alive, sleeping and waking. A cognitive view on metaphors in language use.* Since 2000 she is co-editor of the journal Gesture and co-editor of two edited volumes: with Roland Posner *The semantics and pragmatics of everyday gestures* (2001); with Alan Cienki *Metaphor and Gesture* (in prep.). Current research interests are linguistic analyses of co-speech gestures, cognition and language use, multi-modal metaphors, methods in gesture analysis.

Patricia Osborne and **Esther Olshansky** are Ph. D. students at York University.

Asli Özyürek, Ph.D. in Linguistics and Psychology in 2000 from the University of Chicago. Currently she is an Assistant Professor in Linguistics at Radboud University and a research associate at the Max Planck Institute for Psycholinguistics in Nijmegen. She does research on relations between speech and gesture in production and comprehension as well as on sign languages and gesture systems of "homesigner" children. She is also interested in the relations between language and conceptualization and what gestures, sign languages and homesign systems reveal about this relation.

Simone Pika, Ph.D. in Biology in 2003 from Westfälische Wilhelms University Münster, Germany. Worked at the MPI for Evolutionary Anthropology in Leipzig from 1999–2003. She conducted her postdoctoral research at the University of Albert, Canada and the University of St. Andrews, Scotland. Currently she is a lecturer at the School of Psychological Sciences, Manchester. Her research interest centres on the development and use of communicative signals of non-human and human primates with a special focus on processes of social cognition and the evolutionary roots of spoken language.

Elena Pizzuto, researcher of the Italian National Research Council (CNR), currently coordinates the Sign Language Laboratory at the CNR Institute of Cognitive Sciences and Technologies. Her research focuses on the linguistic investigation of Italian Sign Language (LIS) in a crosslinguistic, crosscultural perspective, and on language development in hearing and deaf children.

Alice Catherine Roy did a Ph.D. in Neuropsychology, on the motor control of reach and grasp in monkeys. During her post-doctoral fellowships in Giacomo Rizzolatti's lab in Parma, and in Luciano Fadiga's lab in Ferrara, she addressed the issue of the link between speech and motor control in humans. Researcher in the Institute of Cognitive Sciences, CNRS, she is now investigating the relation that may exist between syntax and distal motor control.

Michael Tomasello, Ph.D. in Psychology in 1980 from University of Georgia; taught at Emory University and worked at Yerkes Primate Center from 1980 to 1998; since 1998, Co-Director, Max Planck Institute for Evolutionary Anthropology, Leipzig. Research interests focus on processes of social cognition, social learning, and communication in human children and great apes. Books include *Primate Cognition* (w/J. Call, Oxford University Press, 1997), *The New Psychology of Language: Cognitive and Functional Approaches to Language Structure* (edited, Erlbaum, 1998), *The Cultural Origins of Human Cognition* (Harvard University Press, 1999), *Constructing a Language: A Usage-Based Theory of Language Acquisition* (Harvard Universuty Press, 2003).

Grace Vitale is currently a contract faculty member in the Psychology department at York University.

Virginia Volterra. Since 1977 she held the position of Research Scientist and, subsequently, Research Director of the Italian National Research Council (CNR). From 1999 to 2002 she directed the CNR Institute of Psychology (now Institute of Cognitive Sciences and Tecnologies). Her research has focused on the acquisition and development of language in children with typical and atypical development (cognitive impairments and/or sensory deficits) and she has conducted pioneering studies on Italian Sign Language, the visual-gestural language of the Italian Deaf community. She is the author or co-author of over 150 national and international publications in several fields: linguistics, psycholinguistics, developmental psychology, and neuropsychology.

Introduction

Gestural communication in nonhuman and human primates

Katja Liebal, Cornelia Müller, and Simone Pika
University of Portsmouth / European University Viadrina / University of Manchester

What is a gesture?

To answer this question might be as difficult as to describe the concept of time in a few sentences. Researchers have looked at gestures using a variety of research questions and methodological approaches, as well as different definitions. The majority of studies investigated gestures in humans, but recent research started to include different species of non-human primates, particularly great apes but also monkeys. To enable an intense discourse and an interdisciplinary, comparative exchange between researchers interested in different fields of gesture research, a workshop on "Gestural communication in nonhuman and human primates" was held at the Max Planck Institute for Evolutionary Anthropology in Leipzig, March 2004. This multidisciplinary perspective is essential to explore such fundamental questions as the evolution of language as well as the phenomenon of gesture as such: the multiple facets of cognitive, affective, and social functions of gestures, their forms of uses, their varying structural properties, and the cognitive processes such as intention and abstraction involved in the creation and use of gestural signs. Studying gestures in nonhuman and human primates appears therefore a highly interesting enterprise; not only because of their shared phylogenetic history but because of their close relation to language. Gesture is the modality which may take over the burden of vocal language if needed for physiological or ritual reasons (as in sign languages of the deaf and in alternate sign languages (Kendon, 1988, 2004; Senghas, Kita, and Özyürek, 2004). In other words, gestures may develop into a full fledged language under certain conditions. Taking this potential seriously may help to throw new light on the hypothesis that gesture might have been the modality which contributed to the evolution of vocal language in one or the other way.

Hence it seems that a comparative approach would profit significantly from the clarification of some fundamental issues such as how a gesture can be defined and where intentionality does come into play. In addition, it is necessary to answer the question about the extent to which gestural communication systems of nonhuman and human primates are comparable and which methodological steps are essential with regard to data collection, analysis and coding to enable an appropriate comparison of results across species. There is a need for a comparison of structural properties of gestures, to differentiate and classify different kinds of gestures and to investigate the functional contexts in which they are used, to describe the semiotic structures of gestures and how they relate to cognitive processes. Answering these questions will help to clarify whether and how comparative studies of gestural communication in nonhuman and human primates contribute to the question of a likely scenario of the evolution of human language.

This volume consists of four parts and covers a broad range of different strands in the study of gestures. It summarizes the majority of presentations of the workshop on "Gestural communication of nonhuman and human primates", but also includes some additional articles.

Part I provides a theoretical framework of the evolution of language assuming a gestural origin. Part II is focused on research in gestures of nonhuman primates including sign language-trained great apes. Part III addresses gestural communication in humans, such as gesture use in preverbal children and during early language acquisition, speech-accompanying gestures in adults and gestures used by a special population of deaf children, so called 'home-signers'. Part IV explores the potential of a comparative approach to gestural communication and its contribution to the question of the evolution of language.

In Part I, Alice Roy and Michael Arbib offer new arguments for a neurobiologically grounded theory of a gestural origin of human language. The authors further develop the mirror system hypothesis (Arbib, 2005a, b; Arbib & Rizzolatti, 1997; Rizzolatti & Arbib, 1998), i.e. which assumes that the mirror system – supporting production and perception of hand movements in humans and nonhumans – might have played a critical role in the evolution of language and therefore provide a highly pertinent theoretical frame for an evolutionary scenario including a gestural origin of human language.

Part II concerns gestural communication of nonhuman primates, starting with a chapter by Simone Pika, Katja Liebal, Josep Call, and Michael Tomasello. The authors aim at a systematic investigation of the gestural repertoire and its use in gibbons and great apes focusing on how the different ecological, social, and cognitive conditions might influence the respective characteristics of the species' different gestural repertoires. The chapter by Juan Carlos Gomez provides an overview of studies focusing on begging behavior of captive monkeys and apes. It discusses whether request gestures are simply conditioned responses or whether they serve as primitive referential signals based upon a causal understanding of the attentional contact and direction. The next two chapters present empirical studies on gestures used by monkeys and great apes. Dario Maestripieri describes the impact of the social organization on the frequency and contextual use of gestures in three macaque species each of them realizing a different social system. David Leavens and William D. Hopkins report how food size and distance influence the communicative behavior of chimpanzees during interactions with humans including manual gestures, but also gazing and vocalizations. The last chapter of the first part by Valerie Chalcraft and Allen Gardner concerns the use of sign language by chimpanzees. It shows that chimpanzees – as human signers – directionally modulate signs to indicate actor and instrument but also quantitatively modulate signs to indicate intensity.

Part III presents studies on gestural communication in humans. Ulf Liszkowski provides an overview of communicative and social-cognitive abilities of preverbal infants and relates these studies to recent findings on pointing in twelve-months old children. Different aspects of the relationship between gesture and language in early language acquisition are the topics of the three following chapters. Olga Capirci, Annarita Contaldo, Cristina Caselli, and Virginia Volterra focus on gesture use in Italian children between the age of 10 and 23 months. Elena Pizzuto and Micaela Capobianco describe the use and interaction of both deictic and representational elements in Italian children's early gestural-vocal system. Joanna Blake, Grace Vitale, Patricia Osborne, and Esther Olshansky report on a cross-cultural comparison of gestures in human infants during the transition to language between 9 and 15 months of age.

Two chapters focus on gesture in relation to speech and language – including the relation of gesture to a signed language. Asli Özyürek, Sotaro Kita, Shanley Allen, Reyhan Furman, and Amanda Brown show that the linguistic framing of events influences co-speech gestures of adult Turkish and English speakers. Susan Goldin-Meadow describes that gestures may take different forms depending on whether they are produced with speech (gestures as parts of language) or without speech (gestures as language) referring to speech-accompanying gestures on the one hand as opposed to the 'home signs' of deaf kids. This highlights the linguistic potential of gesture if vocal language is not available. The variety of aspects of gestural communication presented in this volume indicates that there is quite some ground to cover for further comparative studies of nonhuman and human forms of gestural communication. Therefore, in part IV, Cornelia Müller seeks to spell out this potential in a more systematic way by taking up the framing questions of the workshop and exploring why a comparative view might offer interesting insights both for researchers of nonhuman and human primates, how comparative studies may further contribute to the dispute over the evolution of language, and what are fundamental conceptual and methodological prerequisites for future comparative research. This chapter therefore offers a condensed presentation of the purpose of this volume: it indicates the current state of the art in the study of gestural communication in nonhuman and human primates and aims at stimulating further interdisciplinary and comparative studies of a wide variety of primate species including humans.

Acknowledgement

We would like to thank Fritz-Thyssen-Stiftung (www.fritz-thyssen-stiftung.de) for funding the workshop on "Gestural communication in nonhuman and human primates" in Leipzig, 2004.

References

Arbib, M. A. (2005a). From monkey-like action recognition to human language: An evolutionary framework for neurolinguistics. *Behavioral and Brain Sciences, 28*(2), 105–124.

Arbib, M. A. (2005b). Interweaving Protosign and Protospeech: Further Developments Beyond the Mirror. *Interaction Studies, 6*(2), 145–171.

Arbib, Michael & Giacomo Rizzolatti (1997). Neural expectations: A possible evolutionary path from manual skills to language. *Communication and Cognition*, 29, 393–424.

Kendon, Adam (1988). How gestures can become like words. In Fernando Poyatos (Ed.) *Cross-cultural perspectives in nonverbal communication* (pp. 131–141). Toronto: C. J. Hogrefe, Publishers.

Kendon, Adam (2004). *Gesture: Visible action as utterance.* New York: Cambridge University Press.

King, B.J. (1999). *The evolution of language: Assessing the evidence from nonhuman primates.* Santa Fe: School of American Research.

Rizzolatti, Giacomo & Michael A. Arbib (1998). Language within our grasp. *Trends in Neurosciences*, 21, 188–194.

Senghas, A., Kita, S., & Özyürek, A. (2004). Children Creating Core Properties of Langauge: Evidence from an Emerging Sign Language in Nicaragua. *Science, 305*, 1779–1782.

Part I

Evolution of language and the role of gestural communication

The syntactic motor system

Alice C. Roy and Michael A. Arbib
Università di Ferrara / Computer Science, Neuroscience and USC Brain Project, Los Angeles

The human brain has mechanisms that can support production and perception of language. We ground the evolution of these mechanisms in primate systems that support manual dexterity, especially the mirror system that integrates execution and observation of hand movements. We relate the motor theory of speech perception to the mirror system hypothesis for language and evolution; explore links between manual actions and speech; contrast "language" in apes with language in humans; show in what sense the "syntax" implemented in Broca's area is a "motor syntax" far more general than the syntax of linguistics; and relate communicative goals to sentential form.

Introduction

Much of the current debate on language evolution consists of establishing whether or not language in general and syntactic processes in particular have emerged on their own or as a by-product of other cognitive functions (Hauser et al., 2002). As to the latter hypothesis, the most influential proposition is the determinant role of motor control in the origin of language. If syntax is in some way a "side effect" of the evolution of the motor system, then syntax might share its cortical territories with this original function (Bates & Dick, 2002) or at least involve adjacent territories which emerged as evolution expanded capacities for imitation, increased flexibility in symbolization and shaped resources for new modes of phonological articulatory control. Here we will briefly review the literature concerning the motor origin of language with special emphasis on the Mirror System Hypothesis, examine the problem of the uniqueness of syntax and finally discuss the possibility of gestures and praxis providing a "syntactic motor system" that is the precursor of the syntactic system of language.

The mirror system and the motor theory of speech perception

In the seventies, the linguist A. Liberman and his colleagues proposed a new view of language acquisition and comprehension. The main postulate of his theory, known as the motor theory of speech perception, was that the core signal emitted and perceived in speech is not the sound so much as the articulatory movements which produce it (Liberman & Mattingly, 1985; Liberman & Whalen, 2000). The authors posited that the articulatory movements are directly perceived as phonetic elements without the need of a cognitive translation; this affirmation automatically posits that the link between phonetics and gestures is not learned throughout association but is instead innate. We would not go so far as to exclude cognitive translation, but would rather see the cognitive as rooted in motor representations and the perceptual structures that access them. Indeed, Studdert-Kennedy (2002) refined the motor theory by adding the proposal that action production and recognition are the key to access to the symbolic order, proceeding through manual action imitation, facial action imitation and then vocal imitation.

The idea of a tight link between language and motor control is indeed much older, and can be traced back to the work of Bonnot de Condillac (1715–1780), a French philosopher who suggested that "a natural language" has been progressively transformed into "a language of action". A fearful scream triggered by the presence of a predator (natural language), for example, could have been associated with the presence of a predator and then reproduced out of its natural context to evoke in someone else's brain the mental image of the predator (language of action). Later on, we shall be somewhat more rigorous in the use of the term "language". In any case, the cornerstone of this theory is the hypothesis that one could recognize the action of the other as a part of one's own motor repertoire to access its meaning. Amazingly, Bonnot de Condillac developed his theory in Parma (Falkenstein, 2002), the place where, some centuries after, the potential neurobiological basis of his theory, the mirror neuron system, was discovered (di Pellegrino et al., 1992). Indeed, in area F5, the rostral part of the macaque's ventral premotor cortex (homologous to Broca's area in the human brain) a new class of neurons was identified by Rizzolatti and his colleagues in Parma (Gallese et al., 1996; Rizzolatti et al., 1996; Umilta et al., 2001). The peculiarity of these neurons is that they discharge both when the monkey executes a specific manual action and when he observes another primate (human or non-human) executing the same specific action, as if they were "recognizing" the aim of the action.

The actions that trigger mirror neurons are transitive, i.e., they involve the action of a hand upon an object, not a movement of the hand in isolation. Moreover, mirror neurons show congruence between the motor action they code and the visual actions they respond to, so that a neuron coding for the whole hand prehension will be preferentially triggered by the observation of the same type of prehension as opposed to another one (e.g., precision grip).The term "resonance" has been used to describe this "mirror property", reflecting the way in which one guitar string may vibrate in response to the vibration of another at a critical frequency. However, where such auditory resonance is a direct physical phenomenon, the "resonance" of a mirror neuron no doubt reflects the result of neural plasticity shaping the neuron to respond to neural codes for visual or auditory patterns associated with its related actions.

The Mirror System Hypothesis

The Mirror System Hypothesis (Arbib & Rizzolatti, 1997; Rizzolatti & Arbib, 1998) asserts that the *parity requirement* for language in humans — that what counts for the speaker (or signer) must count approximately the same for the hearer (or observer) — is met because Broca's area (often thought of as being involved primarily in speech production) evolved atop the mirror system for grasping with its capacity to generate and recognize a set of actions. However (as we shall discuss further below) one must distinguish the mirror system for the sign (phonological form) from the neural schema for the signified, and note the need for linkage of the two. One should also note that, although the original formulation of the Mirror System Hypothesis was Broca's-centric, Arbib and Bota (2003) stress that interactions between parietal, temporal, and premotor areas in the monkey brain provide an evolutionary basis for the integration of Wernicke's area, STS and Broca's area in the human. On this view, Broca's area becomes the meeting place for phonological perception and production, but other areas are required to link phonological form to semantic form.

In any case, the Mirror System Hypothesis provides a neural basis for the claim that hand movements grounded the evolution of language. Arbib (2002, 2005a) modified and developed the Rizzolatti-Arbib argument to hypothesize seven stages in the evolution of language, with imitation of grasping grounding two of the stages. However, as we discuss in this article, research in neurophysiology has given us new insights into macaque neurons in F5 that are responsive to auditory stimuli or are tuned for oro-facial gestures.

The first three stages presented in Arbib (2002) are pre-hominid:

S1: Grasping.
S2: A mirror system for grasping, shared with the common ancestor of human and monkey.

S3: A simple imitation system for grasping, shared with the common ancestor of human and chimpanzee. Here, simple imitation is the ability to acquire some approximation to a movement after observing and attempting its repetition many times.

The next three stages then distinguish the hominid line from that of the great apes:

S4: A complex imitation system for grasping. Here, *complex imitation* combines the ability to recognize another's performance as a set of familiar movements with the ability to use this recognition to repeat the performance, and (more generally) to recognize that another's performance combines novel actions which can be approximated by (i.e., more or less crudely be imitated by) variants of actions already in the repertoire and attempt to approximate it on this basis, with increasing practice yielding increasing skill.

S5: *Protosign*, a manual-based communication system, breaking through the fixed repertoire of primate vocalizations to yield an open repertoire. This involves the breakthrough from employing manual actions for praxis to making such actions exclusively for communication, extending to the repertoire of manual actions to include pantomime of non-manual actions, and then going beyond pantomime to ritualize certain of its performances and add conventionalized gestures that can disambiguate pantomimes (e.g., modifying a single pantomime to distinguish [at least] the three meanings of "bird", "flying" and "bird flying").

S6: *Protospeech*, resulting from the ability of control mechanisms evolved for protosign to link with a vocal apparatus of increasing flexibility.

The hypothesis is that protosign built up vocabulary by variations on moving handshapes along specific trajectories to meaningful locations; whereas protospeech "went particulate". Arbib (2005b) argues that we should not imagine that Stage S5 "went to completion" prior to Stage S6, but rather that protosign and protospeech evolved in an expanding spiral.

In our view, these six stages do not (in general) replace capabilities of the ancestral brain so much as they enrich those capabilities by embedding them in a more powerful system. The final stage is then:

S7: *Language*: the change from action-object frames to verb-argument structures to syntax and semantics.

It is still controversial whether Stage S7 resulted from historical changes in *Homo sapiens* rather than biological evolution beyond that needed for Stages S1–S6 (Arbib, 2002, 2005a), or instead whether the emergence of syntax as we know it in language required further neurobiological evolution to support it. The present article makes two contributions to this argument by (a) beginning to chart the extent to which manual behavior does and does not have a syntax in the sense in which language does; and (b) providing mechanisms which may have made possible the essential contributions that Stage S5, protosign, is claimed to have made to Stage S6, the emergence of protospeech.

The Saussurean sign

Figure 1 makes explicit the crucial point (Hurford, 2004), noted earlier, that we must (in the spirit of Saussure) distinguish the "sign" from the "signified". In the figure, we distinguish the "neural representation of the sign" (top row) from the "neural representation of the signified" (bottom row). The top row of the figure makes explicit the end result of the progression of mirror systems, described in the previous section, from grasping and manual praxic actions via various intermediate stages (Arbib, 2002, 2005a) to conventionalized manual, facial and vocal communicative gestures — to what we will, for the moment, call "words".

The bottom row is based on schema theory (Arbib, 1981, 2003), which distinguishes *perceptual schemas* which determine whether a given "domain of interaction" is present in the environment and provide parameters concerning the current relationship of the organism with that domain, and *motor schemas* which provide the control systems which can be coordinated to effect a wide variety of actions. Recognizing an object (a candle, say) may be linked to many different courses of action (to place the candle in one's shopping basket; to place the candle in a drawer at home; to light the candle; to blow out the candle; to choose a candle among several, etc.). In this list, some items are candle-specific whereas other invoke generic

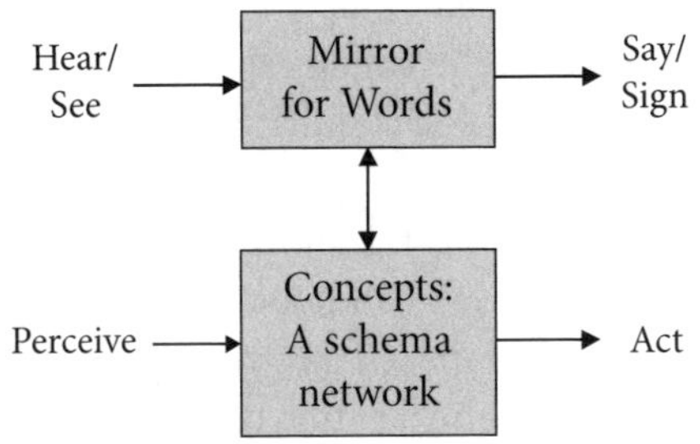

Figure 1. The bidirectional sign relation links words and concepts. The top row concerns Phonological Form which may relate to signed language as much as to spoken language. The bottom row concerns Cognitive Form and includes the recognition of objects and actions (Arbib, 2004, after Hurford, 2004).

schemas for reaching and grasping. Only for certain basic actions, or certain expressions of emotion, will the perceptual and motor schemas be integrated into a "mirror schema". A "concept" does not correspond to a unique word, but rather to a graded set of activations of the schema network.

As a result, the form of the Mirror Property posited for communication — that what counts for the sender must count for the receiver — does not result from the evolution of the F5 mirror system *in and of itself* to support communicative gestures as well as praxic actions. It is also crucial that the evolution of neural codes for these communicative actions occurs within the neural context that links the execution and observation of an action to the creature's planning of its own actions and interpretations of the actions of others. These linkages extract more or less coherent patterns from the creature's experience of the effects of its own actions as well as the consequences of actions by others to provide meaning to the communicative actions correlated with the action and much that defines its context. Similarly, execution and observation of a communicative action must be linked to the creature's planning and interpretations of communication with others in relation to the ongoing behaviors which provide the significance of the communicative gestures (compare and contrast the language of action of Bonnot de Condillac).

In speech, a word consists of a sequence of "phonemes", and although the division into phonemes is somewhat artificial, the key point is that the underlying reality is the concurrent movement of a range of articulators (though some such actions may be mono-articular; cf. Studdert-Kennedy, 2002). Similarly, turning to manual control, arm movements generally involve shaping the hand and use of the hand needs a reach to position it properly. Within the context of the Mirror System Hypothesis, this raises the issue of whether a "reach and grasp" is more like a word or a phoneme. Our answer is — paradoxically — both (Arbib, 2005c). We earlier suggested that a key point in the evolution of brain mechanisms underlying language (Stage S4, complex imitation) involved the ability to recognize that a novel action was in fact composed of (approximations to) known actions. This recognition is not only crucial to the child's capacity for "complex imitation" and the ability to acquire language and social skills, but is also essential to the adult use of language. In both signed language and speech, we recognize a novel utterance as in fact composed of (approximations to) known actions (namely the speaking or signing of words) and, just as crucially, the stock of words is open-ended.

However, signed language and speech take very different approaches to the formation of words. Signing exploits the fact that the signer has a very rich repertoire of arm, hand and face movements, and thus builds up vocabulary by variations on the multi-dimensional theme "move a handshape [or two] along a trajectory to a particular position while making appropriate facial gestures". By contrast, speech employs a system of vocal articulators which have no rich behavioral repertoire of non-speech movements to build upon. Instead speech "went particulate", so that the spoken word is built (to a first approximation) from a language-specific stock of phonemes (actions defined by the coordinated movement of several articulators, but with only the goal of "sounding right" rather than conveying meanings in themselves). In summary, a basic "reach and grasp" corresponds directly to a single "word" in signed language; whereas in speech, a basic "reach and grasp" is more like a phoneme, with a word being one level up the hierarchy.

But if single actions are the equivalent of phonemes in speech or words in sign, what levels of motor organization correspond to derived words, compound words, phrases, sentences, and discourse; what motor control levels could there possibly be at these sequentially more inclusive levels? Getting to derived words seems simple enough. In speech, we play variations on a word by changing speed and intonation, and by various morphological changes which may modify internal phonemes or add new ones. In sign, "words" can be modified by changing the source and origin, and by various modifications to the path between. For everything else, it seems enough — for both action and language — that we can create hierarchical structures subject to a set of transformations from those already in the repertoire. The point is that the brain must provide a computational medium in which already available elements can be composed to form new ones, irrespective of whether these elements are familiar or not. It is then a "cultural fact" that when we start with words as the elements, we may end up with compound words or phrases, other operations build from both words and phrases to yield new phrases or sentences, etc., and so on recursively. Similarly, we may learn arbitrarily many new motor skills based on those with which we are already familiar. There seems to be no more (or less) of a problem here for motor control than for language. We form new words by concatenating phonemes in speech, and by combining handshapes and trajectories in sign. Once we get to the word level, we proceed similarly (but with different details of the syntax) in the two cases.

However, having emphasized the differences in "motoric level" between the words of signed language and speech, we now show that there is nonetheless a tight linkage between the modalities of manual actions and speech.

Manual actions and speech: The origin of the link

According to the mirror system hypothesis, hand movements have played a determinant role in the emergence of a representational system enabling communication, with a mirror system underwriting the parity of speech and perception. The (updated) motor theory of speech perception places vocalization within a similar framework. We now review a number of clues suggestive of a close biological link between these manual and vocal systems. First, it is important to recall that performing and controlling fine manual actions is not a trivial task and that primates are precocious among the animal kingdom in their mastery of this ability. Drawing a parallel, humans are unique among primates in possessing the faculty of language. Birds have superb vocal control without manual skill but their vocalizations were not sufficient to develop language; conversely the development of sign languages by the deaf community shows that humans are able to develop an autonomous language in the absence of vocalizations. Thus, as we seek to understand the special nature of the human speech system within the evolutionary context afforded by the study of primates, we argue that manual dexterity provides a key to understanding human vocalization.

While the presence of right-handedness in apes is still a matter of debate (Corballis, 2003; Hopkins & Leavens, 1998; McGrew & Marchant, 2001; Palmer, 2002), it is well known that on average 90% of the human population is right handed, having in general the left hemisphere in charge of controlling the distal musculature. Since the seminal work of Broca, we know that language is also implemented in the left hemisphere and indeed cerebral asymmetry for language and handedness are correlated (Knecht et al., 2000; Szaflarski et al., 2002). Rather than an exact rule (as 78% of left-handers still present a language dominance in the left hemisphere and 7.5% of right-handers present the opposite lateralization for language) this indicates that the dominance for hand dexterity and language may not be casual and that evolution at a certain point may have favored this type of organization. In the organization of cerebral cortex, mouth (and face in general) and hand somatomotor representations are contiguous, leading in particular pathological cases to a functional overlap (Farnè, et al., 2002). In this study, the authors examined the performance of a patient who had benefited from having both a left and right hand allo-grafted to report single and double simultaneous tactile stimulations. Five months after the surgery, the patient was perfectly able to report single stimulation to the grafted hands. However in the case of a double simultaneous stimulation delivered on the right hand and right jaw, the patient's performance dropped dramatically, as in half of the trials a tactile sensation was reported only on the jaw. The absence of this facial-manual overlap under the same conditions six months later clearly indicates that the cortical reorganization and competition between the territories of the hand and the face that occurred after the amputation and the graft were responsible for the initial functional overlap (Giraux et al., 2001).

In hand-reared chimpanzees, fine motor manipulations are often accompanied by mouth and tongue movements (Waters & Fouts, 2002). Moreover, both hand and mouth are prehensile organs, as is well observable in newborns. In sum, cortical representations and functions of hand and mouth are so intricately interwoven that finally it is not so surprising to observe that blind people gesticulate when speaking even though they can see neither their visual gestures nor their effects on others (Iverson & Goldin-Meadow, 1998). However, it might be erroneous to restrict the link between speech and hand movements to a low-order factor such as motor co-activation. In humans, speech is the most common way to communicate, but sign language substitutes perfectly for speech in the deaf.

A step closer to language

Since its discovery, our knowledge about the mirror system has increased considerably. In monkeys, discussion of the possible link between the mirror system and the origin of language has been enriched by the discovery of acoustic mirror neurons (Kohler et al., 2002). Acoustic mirror neurons discharge both when the animal performs a specific manual action which is associated with a characteristic sound (e.g., breaking a peanut), and when the monkey sees the same action performed by someone else or only hears the sound that is produced by the action. These multisensory mirror neurons make possible the link between a heard sound and the action that produces it. This is somewhat akin to the link proposed by Liberman and Whalen (2000), though their theory emphasizes articulatory movements during speech production and hearing rather than the concomitants of manual actions. Until recently, mirror neurons had been observed only for hand actions, leaving the gap between hand movement recognition and recognition of vocal articulatory movements unfilled. More recently, mouth mirror neurons have been identified in monkey ventral premotor cortex (Ferrari et al., 2003). Two types of mouth mirror neurons have been described. Neurons of the first class are active during executed and seen ingestive behaviors. Those of the second class respond to communicative gestures (e.g., lip smacking) and thus provide additional evidence in favor of a fundamental role of mirror neurons in the emergence of language (Rizzolatti & Arbib, 1998). This is not to say that the mirror system of monkeys is already developed enough to provide a language-ready brain, nor to support the view of the evolution of language as being primarily vocal without the key involvement of manual dexterity. The new classes of mirror neurons are rather to be seen as some of the primary colors a painter needs to be able to create all the nuances of the palette.

We must here add a fundamental piece of evidence: the existence of a mirror system in humans. In the last decade, brain imaging studies as well as Transcranial

Magnetic Stimulation (TMS)[1] studies have consistently demonstrated the existence in the inferior frontal gyrus of what can be interpreted as a human mirror system for hand actions (Buccino et al., 2001; Grèzes et al., 2003; Fadiga et al., 1995). The inferior frontal gyrus corresponds to Broca's area (BA 44–45), a cortical area which diverse studies have related both to the motor system and language functions. It has been thus put forward that Broca's area in humans might be the functional homologue of area F5 of monkey's premotor cortex. Several lines of evidence support this view. Among these, Fadiga and coworkers (2002) have shed light on the motor resonance occurring in correspondence to listening to words. They demonstrated that the tongue motor evoked potentials reached higher amplitudes when their (Italian) subjects were listening to Italian words that recruited important tongue movements (birra) when compared to words recruiting less important tongue movements (buffo). The functional role of such a peculiar phenomenon can be explained readily in an expanded view of Liberman's motor theory of speech perception in which the circuitry for language sound recognition is bound up with a mirror system for the generation and recognition of mouth articulatory movements. On this account, recognition of mouth articulatory movements should be embedded in the heard representation of a word.

More recently, Gentilucci, Roy and colleagues, using a behavioral approach, have investigated the tight link between manual actions and speech production. We see this work as supporting the mirror system hypothesis for the evolution of language by showing that manual gestures relevant to communication could have natural vocal concomitants that may have helped the further development of intentional vocal communication. In a first study, we (Gentilucci, Santunione, Roy, & Stefanini, 2004) asked each subject to bring a fruit of varying size (a cherry or an apple) to the mouth and pronounce a syllable instead of biting the fruit. We found an effect of the fruit size not only on the kinematics pattern of the mouth aperture but also and more importantly on the vocal emission of the subjects. By analyzing the vocal spectrum it emerged that formant 2 (F2) was actually higher when bringing the large fruit rather than the small one to the mouth. F2 is, like the other formants, an acoustic property of the vocal tract that produced the spectrum. The frequency of F2 is known to be tightly linked to the shape of the vocal tract. Our experiment demonstrated that the fruit size influenced the vocal tract configuration which in turn modified the frequency of F2. The effect observed was present also when subjects pronounced the syllable when just observing, without executing, the same arm action being performed or pantomimed by someone else.

While this study highlights the potential role of upper limb action and the underlying mirror system mechanisms in the emergence of vocal signs, a second study goes further by revealing the specificity of the link between manual action and vocal emission (Gentilucci, Stefanini, Roy, & Santunione, 2004). In this case, we asked subjects to observe two types of manual action, a bringing to the mouth action and a prehension movement. In each case, the action was performed with a small or a large fruit and the subjects had to pronounce the syllable at the end of the movement. The vocal parameters affected by the fruit size changed according to the type of movement observed. While the second formant varied during the bringing to the mouth task, the first formant varied during the prehension task. Our results are of particular interest as they suggest that the emergence of voice modulation and thus of an articulatory movement repertoire could have been associated with, or even prompted by, the preexisting manual action repertoire.

Finally, we note that McNeill and Goldin-Meadow found that manual co-speech gestures[2] may convey additional information and thus complete speech (McNeill, 1992; Goldin-Meadow, 1999). Moreover, the production of co-speech gestures by blind persons talking to each other indicates how ancestral is the link between hand and language (Iverson & Goldin-Meadow, 1998). Indeed, evidence of a linkage between manual skills and vocalization has been reported in macaques by Hihara, Yamada, Iriki, and Okanoya (2003). They trained two Japanese monkeys to use a rake-shaped tool to retrieve distant food. After training, the monkeys spontaneously began vocalizing coo-calls in the tool-using context. Hihara et al. then trained one of the monkeys to vocalize to request food or the tool:

Condition 1: When the monkey produced a coo-call (call A), the experimenter put a food reward on the table, but out of his reach. When the monkey again vocalized a coo-call (call B), the experimenter presented the tool within his reach. The monkey was then able to retrieve the food using the tool.

Condition 2: Here the tool was initially presented within the monkey's reach on the table. When the monkey vocalized a coo-call (call C), the experimenter set a food reward within reach of the tool.

The intriguing fact is that the the monkey spontaneously differentiated its coo-calls to ask for either food or tool during the course of this training, i.e., coos A and C were similar to each other but different from call B. Hihara et al. speculate that this process might involve a change from emotional vocalizations into intentionally controlled ones by associating them with consciously planned tool use. However, we would simply see it as an example of the unconscious linkage between limb movement and vocal articulation demonstrated in humans by Gentilucci, Roy and their colleagues.

Fundamentals of "language" in apes

As we share 98.8% of our DNA with our closest relative, the chimpanzee (Fujiyama et al., 2002), it is of interest to track the extent to which language has appeared in apes. The quotes around "language" in the title of this section is to highlight the fact that nonhuman primate communication is very different from human language, and that even apes raised by humans can develop only a small vocabulary and seem incapable of mastering syntax. Two main streams of research can be dissociated, of which the first tried to teach language to apes while the second mainly observed the communicative gestures used in ape communities without human instruction.

Attempts to teach apes to talk failed repeatedly (Kellog & Kellog, 1933; Hayes, 1951), though comprehension of spoken words has been demonstrated by apes. First, apes are limited in their capacity to emit vowels by the anatomical configuration of the larynx (Nishimura et al., 2003). Second, vocalizations in apes primarily serve emotional functions, their capacity for modulating vocalizations voluntarily being still debated (Deacon, 1997; Ghazanfar & Hauser, 1999). Kanzi, the most famous bonobo, is able to understand 400 spoken English words, but his understanding of syntax is almost non-existent, having been compared to that of a two year old child (Savage-Rumbaugh et al., 1998). Moreover, Kanzi's comprehension was impaired when the same word conveyed different meanings in a single sentence (e.g., *Can* you use the *can*-opener to open a *can* of coke?). Moreover, we should mention that Kanzi seemed to be particularly smart — other trained apes never reached his level of language ability.

A more successful approach has focused on the use of the hands, teaching apes the use of hand signs like those used in sign language[3] or the placing of visual symbols called lexigrams (Gardner & Gardner, 1969; Savage-Rumbaugh et al., 1998). These complement two types of communicative gestures seen in apes. The first type is naturally present in the repertoire of the species in the wild, the other appears in apes raised by humans or at least in extensive contact with humans (Tomasello & Call, 2004):

The first type compounds manual and bodily gestures "that are used to get another individual to help in attaining a goal" (Pika et al., 2005) and mainly take place in functional contexts such as play, grooming, nursing, and during agonistic and sexual encounters (Pika et al., 2003, 2005; Liebal et al., 2004).

The other type consists of gestures mostly performed during interactions with humans and often used to request food (see Gomez this issue). Pointing gestures for example seem to be "human dependent" as pointing has been observed only once in wild bonobos (Vea & Sabater-Pi, 1998). In captive conditions, chimpanzees have been seen to develop pointing gestures that can be directed to other congeners or human beings, without being taught pointing movements (Call & Tomasello, 1994; Leavens & Hopkins, 1998; 1999; Hopkins & Leavens, 1998). The discordance of behavior between wild and captive chimpanzees can find an explanation in the impossibility, for captive chimpanzees, of reaching directly for the object of their interest, being obliged to develop deictic pointing gestures to signify their need to a mediator (a human or a congener) who is closer to the object or can move toward it. This hypothesis finds support in the observation that pointing in human babies occurs primarily towards targets which are clearly out of reach (Butterworth, 2003). The particular immature state of the locomotion system of humans at birth may have driven the species to develop a deictic pointing behavior. Moreover, chimps accompanied their deictic gestures with eye contact and even vocalizations to capture the attention of the audience (Leavens, 2003; Povinelli et al., 2003). However, chimpanzees, like human babies, use their gesture imperatively (i.e., to get another individual to help in attaining a goal) but not declaratively (as human adults do) to draw another's attention to an object or entity merely for the sake of sharing attention. Gestures in apes are used for dyadic interactions as opposed to a referential use in therefore triadic interactions of pointing gestures in humans (Tomasello, 2006). The ability of apes (but not monkeys) in captivity to produce imperative pointing reveals some form of brain-readiness for a set of communicative gestures beyond those exhibited in the wild. In the same vein, we note that Kanzi learned "language" as human infants do, that is by observing and listening to the "English classes" his mother was taking, rather than being purposely involved in language lessons. This relates to our general view (recall the earlier discussion of Stage S7) that biological substrate and "cultural opportunity" are intertwined in expressing the human readiness for language.

In the context of the mirror system hypothesis, it seems that the most evolved communicative gestures in non-human primates take the shape of deictic movement. Although two studies report iconic gestures in apes (Savage-Rumbaugh et al, 1977 for bonobos; Tanner & Byrne, 1996 for gorillas), the use of these gestures seems to be restricted to single individuals only since these observations have never been replicated for other groups of bonobos (Roth, 1995) or gorillas (Pika et al., 2003). This discrepancy supports the interpretation of Tomasello and Zuberbühler (2002) that "these might simply be normal ritualized gestures with the iconicity being in the eyes of the human only" and that "a role for iconicity [...] has not at this point been demonstrated". Recall the intrinsically transitive nature of the gestures that trigger mirror neurons in the macaque (i.e., they involve the action of a hand upon an object, not a movement of the hand in isolation), and their specificity for one peculiar type of movement rather than another. The impossibility for apes to produce unequivocal iconic gestures that represent a specific action rather than deictic pointing underlines the notion that the adaptation of praxic movements for communicative purposes was indeed an important evolutionary step – the one marked in the mirror system hypothesis by the transition from Stage S4, complex imitation, to Stage S5, protospeech.

Considering now the human infant, deictic gestures can be observed accompanying and even preceding the production of the first word or the first association of two words (Goldin-Meadow & Butcher, 2003), and become enriched by iconic and other meaningful gestures in the co-speech gestures of human beings throughout their lives. Moreover, many of the signs seen in the signed languages used by the deaf seem to have an iconic "etymology" though in fact the neural representation of signed gestures is independent of whether or not the sign resembles a pantomime. Indeed, the neural mechanisms for the signs of language can be dissociated from those for pantomime. Corina et al. (1992) demonstrated the dissociation of pantomime from signing in a lesioned ASL (American Sign Language) signer, while Jane Marshall et al. (2004) described a BSL (British Sign Language) signer for whom gesture production was superior to sign production even when the forms of the signs and gestures were similar. Deaf children, in the absence of any teaching, develop a rudimentary sign language outfitted with a primitive syntax (Goldin-Meadow, 1999, 2002). On the basis of such a primitive syntax, a more complex syntactic system can be progressively developed by new generations of signers, as observed in Nicaraguan deaf children (Senghas et al., 2004) and Sayyid Bedouin Sign Language (Sandler et al., 2005). However, it must be stressed that in each case emergence of the sign language occurred in a community that included speakers of a full human language (Spanish and Arabic, respectively) providing a model of complex communication that could be observed though not heard. In any case, something crucial is lacking in apes that would enable them to fill the gap between scarce communicative gestures and the use of a language of human richness. Syntactic competence seems to be an essential ingredient of language and we thus turn to a discussion of the cortical systems which support syntax in the human brain.

Syntax and Broca's area

The innate versus acquired nature of syntax is the object of a long standing debate which questions whether syntactic rules are innately prestructured (at least partially) or acquired through learning, whether through explicit rules or "rule-like" behavior (Pinker, 1997; Seidenberg, 1997; Albright & Hayes, 2003). The latter distinction is between a view of processing as explicitly invoking a coded representation of rules and one in which a neural network may exhibit patterns of behavior which can be summarized by rules yet with no internal representation of these rules. Discussion of these aspects is outside the scope of the present paper. For the present discussion, rather, we take a more general view, looking not at syntax as a set of rules specific to the grammar of human languages, but rather at syntax more broadly defined as whatever set of processes mediate the hierarchical arrangement of elements governing motor production and thus, in particular (King, 1996), the production of a sentence. We then ask whether syntax in this broader sense can have a cerebral functional localization that, in addition, may give some hints on its possible origins.

In recent years the number of studies aimed at investigating the brain structures involved in syntactic processing has increased dramatically. Brain imaging studies have repeatedly pointed out the crucial role of Broca's area in syntactic processing. While the most anterior part of Broca's area (i.e., BA [Brodmann's area] 45) appears more involved in semantic processing, activation of BA44 has been reported in different languages and in jabberwocky (i.e., sentences in which many of the content words have been replaced by nonsense words, while preserving syntactic markers) during syntactic violation detection, syntactic plausibility judgment, lexical decision and increased syntactic complexity (Hashimoto & Sakai, 2002; Kang et al., 1999; Heim et al., 2003; Embick et al., 2000; Friederici et al., 2000; Moro et al., 2001; Newman et al., 2003). However, BA44 syntactic processing (using our broader sense of the term) does not seem to be limited to language. For example, the sequence of harmonics in much music is predictable. By inserting unexpected harmonics, Maess and coworkers (2001) have studied the neuronal counterpart of hearing musical "syntactic" violations. A bilateral activity of BA44 has been observed, suggesting thus that BA44 is also implicated in rule-like behavior that is not specific to language. Similarly, Broca's area is activated during a compound calculation task, a result suggesting that Broca's area may also be involved in rule-like processing of symbolic information (Gruber et al., 2001).

BA44 has also proven to be important in the motor domain, raising once more the ticklish question of the link between language and the motor system. BA44 is activated during distal movement execution (Lacquaniti et al., 1997; Matsumura et al., 1996; Grafton et al., 1996; Binkofski et al., 1999a, b; Gerardin et al., 2000). The involvement of Broca's area in distal movements rather than in proximal ones is certainly not anecdotal. As we discussed before, distal motor control as manual dexterity is exceptionally developed among primates, whereas control of the more proximal part of the forelimb for reaching is well developed in many mammal species. However, the role of Broca's area in the motor field goes further than simple execution because observation, simulation and imitation of distal and facial movements also strongly involve BA44 activity (Gerardin et al., 2000; Nishitani & Hari, 2000; Iacoboni et al., 1999; Koski et al., 2003; Tanaka & Inui, 2002; Hamzei et al., 2003; Heiser et al., 2003; Grèzes et al., 2003; Carr et al., 2003; Decety & Chaminade, 2003; Leslie et al., 2004). The presence of a goal seems essential, as aimless actions trigger less or no activation of BA44 (Grèzes et al., 1999; Campbell et al., 2001) while the presence of a goal enhanced the activity of BA44 (Koski et al., 2002). At this point it becomes clear that syntactic and distal, fine motor processes share a common neuronal substrate, but why? Did it present a benefit for syntax to develop in a part of the premotor cortex which controls manual actions? Did it succeed by chance or is syntax an emergent property of the motor system?

In search of a "motor syntax"

Looking for homologies between the motor system and the language system, the question of lateralization pops out again. We will not debate here the issues of right handedness and language (see Corballis, 2003) but instead orient our discussion towards the different levels of motor deficits that occur after right or left brain damage. The deficits related to a lesion of motor cortex and premotor cortex affect the contralesional limb, with lesions to either hemisphere inducing similar impairment. The pattern changes considerably if we consider the syndromes following a parietal lobe injury. While neglect, defined as an inability to perceive the contralesional space, generally occurs following a right hemispheric lesion, limb apraxia appears predominantly after a left hemispheric lesion and most often affects both hands (Poizner et al., 1998). Limb apraxia is generally described by exclusion; it is not an impairment attributable to motor weakness, akinesia, intellectual impairment, impaired comprehension or inattention (Harrington & Haaland, 1992). In spite of the different forms of apraxia, a tentative common definition would posit that apraxia is a deficit in the organization of gesture as opposed to movement. While a movement is the motion of a body part, a gesture generally refers to a hierarchically organized sequence of movements directed to a final aim that can be learned (tool use), or convey a meaning (sign of the cross).

In his influential work, Liepmann (1913) identified two high-order types of limb apraxia as evaluated by the class of errors made by patients. Patients suffering from *ideational apraxia* appear unable to construct the idea of the gesture. Ideational apraxics are dramatically impaired in daily life as household tools are no longer associated with specific actions. *Ideomotor apraxia* is more frequent and generally less debilitating. While the *idea* of the movement appears to be preserved, its execution is subject to a voluntary-automatic dissociation. Ideomotor apraxics can present relatively well preserved behaviors as long as they are performed in an ecological context (Schnider et al., 1997; Leiguarda & Marsden, 2000).

The great majority of apraxic patients also suffer from aphasia, an observation suggesting that the neural networks that mediate language and praxis may partly overlap. A double dissociation was only lately reported by Papagno and colleagues (1993). In a cohort of 699 left brain damaged patients they reported 149 cases of aphasia without apraxia and only 10 cases of apraxia without aphasia. Moreover, it has been reported that praxic hemispheric specializations is more closely related to lateralization of language functions than to hand preference (Meador et al., 1999). Here we stress that the "center" for praxis localized in the left hemisphere is responsible for the praxic ability of both the dominant and non-dominant hand. Several aspects of apraxia are of particular interest for us. The striking inability of patients affected by ideomotor apraxia to perform imitation tasks could derive from damage to the mirror system. Mirror neurons in monkeys have been found both in the ventral premotor cortex and in the inferior parietal lobule (Gallese et al., 1996, 2002), two cortical areas that have been linked in human brain imaging study with imitation mechanisms (Koski et al., 2002, 2003, Rizzolatti & Buccino, 2004; Arbib et al., 2000; Rizzolatti et al., 2001; Nishitani & Hari, 2000, Decety & Chaminade, 2003; Rumiati et al., 2004). A recent PET study has qualified this interpretation by demonstrating that the activation in the inferior frontal gyrus was present when the goal of the action was imitated, whereas this activation was no longer present when the means to achieve the goal were imitated (Chaminade et al., 2002). This result fits well with the dissociation observed in apraxic patients between a preserved ability to imitate meaningless gesture and an inability to imitate meaningful gesture (Mehler, 1987).

Another particular aspect of apraxia is that apraxic patients tend to be more impaired for transitive gestures (i.e., those directed toward objects) than for intransitives gestures. The ability to develop and use tools is an important landmark in the cognitive evolution of human species. Here again mirror neurons that respond to tool use have been discovered in monkeys that have been highly exposed to actions made with tools by the experimenter (Ferrari et al., 2005). Tool use is naturally present in apes even if in a small proportion, but it is controversial whether adults coach their offspring (see Boesch & Boesch, 1983; Tomasello, 1999). Apes and humans share a particularity about tool use: the capacity of using different tools for the same end (e.g. using a coin as a screw-driver, Bradshaw, 1997), a capacity no longer present in apraxics (Goldenberg & Hagmann, 1998). However, there is no doubt that humans are unique in the way they can make use of tools (Johnson-Frey, 2003, 2004). While apes and, to some extent, monkeys (Iriki et al., 1996; Hihara, Obayashi, Tanaka, & Iriki, 2003) are able to learn the use of some tools, they lack the critical capacities that enable humans to recognize the need for a tool and thus to create it (Donald, 1999). Bradshaw (1997; see also Beck, 1980) defined a tool as "something used to change the state of another object". The same definition could apply to syntax: a combination of rules that can change the status of a word and, thus, the meaning of a sentence. The problems exhibited by ideational apraxics in structuring their motor acts following a functional hierarchy lead to errors like dialing the number before picking up the receiver or scratching a match on the candle instead of on the box of matches in an attempt to light a candle (Rapcsak et al., 1995). We interpret these deficits as a disruption of a "motor syntactic system". In such a system, each part of the sequence does not so much have a particular order as a particular function or sub-goal (see below) that determines a particular order that enable the "motor sentences" to be performed correctly, that is to maintain its functional goal.

From communicative goal to sentential form

While the operations involved in assigning syntactic structures and using them to determine aspects of meaning may differ from operations in other areas of cognition, we offer a perspective which suggests similarities between action and language which can ground new insights into the underlying neural mechanisms. Consider a conditional, hierarchical motor plan for the just-considered task of lighting a candle with a match:

> While holding a box of matches with the non-dominant hand, use the dominant hand to scratch a match repeatedly against the box until it flares; bring the burning match up to the wick of the candle and wait until it ignites, then move away and shake the dominant hand to extinguish the match.

We have here a hierarchical structure which will unpack to different sequences of action on different occasions. The subsequences of these sequences are not fixed in length a priori, but instead are conditioned on the achievement of goals and subgoals. For example, one may need to strike the match more than once before it flares. We choose this example because its rich intertwining of actions and subgoals seems to us to foreshadow within the realm of action some of the essential ingredients of the syntax of language. There, for example, verb-argument structures express the thematic roles of object with respect to action but various clauses can enrich the meaning of the sentence. One could consider lighting a candle as an action involving an agent (the one whose hand holds the match), a theme (the candle), and an instrument (the match);, the sub-goal of scratching the match then establishes a "clause" within the overall "sentence". In any case, returning to the motor sphere, a "paragraph" or a "discourse" might then correspond to a complex task which involves a number of such "sentences".

Now consider a sentence like

(1) Serve the handsome old man on the left.

spoken by a restaurant manager to a waiter. From a "conventional" linguistic viewpoint, we would appeal to a set of syntactic rules and look for a parse tree whose leaves yield up the word sequence as a well-formed sentence of English. But let us change perspective, and look at the sentence not as a structure to be parsed but rather as the result of the manager's attempt to achieve a *communicative goal*: to get the waiter to serve the intended customer (Arbib, 2006). He could use a *mixed strategy* to achieve his goal, saying "Serve that man." and using a deictic gesture (pointing) to disambiguate which man. However, to develop the analogy with lighting a candle, we consider a *sentence planning strategy* which repeats the "loop" <add adjective or prepositional phrase> until (the manager thinks) ambiguity is resolved:

(1a) Serve the man.

Still ambiguous? Expand it to:

(1b) Serve the man on the left.

Still ambiguous? Expand it to:

(1c) Serve the old man on the left.

Still ambiguous? Expand it to:

(1d = 1) Serve the handsome old man on the left.

Still ambiguous? Apparently not. So the manager says this sentence to the waiter ... but the waiter veers off in the wrong direction. And so the manager says:

(2) No, no. The one who is reading a newspaper.

Note how the error correction is applied without using a whole sentence. The suggestion is that syntactic rules approximated by NP → Adj NP and NP → NP PP (adding an adjective, Adj, or a prepositional phrase, PP, to a noun phrase, NP) can be seen as an abstraction from a set of procedures which serve to reduce ambiguity in reaching a communicative goal.

Clearly, there is no simple map from a set of communicative strategies to the full syntax of any modern language. We believe that any modern language is the result of "bricolage" — a long process of historical accumulation of piecemeal strategies for achieving a wide range of communicative goals. This process of "addition" is complemented by a process of generalization whereby a general strategy comes to replace a range of ad hoc strategies. Thus, just as nouns may serve to denote more than objects, and verbs may serve to denote more than actions, so too do grammatical rules encompass far more than suggested by the simple motivation for NP → Adj NP and NP → NP PP given above.

From phonology to grammar

Research in motor control is more at the level of phonology — how do the effectors produce a basic action, what "co-articulation" may modify one action on the basis of what comes next, etc.? — than at the level of syntax and semantics which analyzes the structure of a full sentence, and, e.g., anaphoric relations between sentences. What we do have is attempts to look at the role of pre-SMA (part of the supplementary motor area) and basal ganglia (BG) and left parietal cortex in fairly simple sequential behavior (Bischoff-Grethe et al., 2003) — which cries out for an understanding of apparent sequences which are better understood as the expression of a hierarchical structure — and then studies of prefrontal cortex (PFC) which discuss planning abilities but tend to be only weakly linked to computational accounts of neural circuitry (Passingham, 1993, Chapter 10). However, one cannot have a fruitful dialogue between the study of the neural basis of action and the study of the neural basis of language unless one accepts that syntax as normally presented is an abstract description, not a process description. The hearer's processes for understanding (more or less) what the speaker intends, and the speaker's processes for conveying the intended message with (more or less) reduced ambiguity must, to be successful, be *approximately* inverse to each other. We may distinguish "production syntax" — getting from a communicative goal to the words that express it — and a "perception syntax" — getting from a sequence of words to the goal behind it. Syntax in the normal sense then is a compact answer to the question: "In this community, what regularities seem to be shared by the sentences that are produced and understood?" In this way, the linguist has some hope of using a *single* grammar to represent regularities which encompass many of the regularities common to both perception and production of utterances — but this does not mean that there is a single grammar represented in the brain in such a way that is consulted by separate processes of perception and production.

By using the candle example to show how actions may need to invoke subactions for their completion, we exhibited the analog of the potential (never actual) infinite regress in the recursive structure of sentences, and used the "identify the customer" example to make more explicit how the language regress might not be as different from the action regress as would seem to be the case if we focus on syntax in the abstract rather than its relation to the forming of the sentence to meet a communicative goal. We do not deny that language does have unique features that separate it from motor planning. The challenges of "linearizing thought" by language are sufficiently different from those of spatial interaction with the world that they may well have required, or given rise to, some specialization of neural circuitry for language. However, at the moment we incline to the view that much of that specialization is due to self-organization of the brain of the child in response to growing up within a language-using community and suggest that language and action both build on the evolutionary breakthrough that gives us a brain able to recognize that a novel gesture is in fact composed of (approximations to) known actions. This ability would consist of extracting syntactic rules and applying them (or the corresponding "rule-like" processors) to known actions/words derived from the mirror system, to recognize and generate new gestures/sentences.

Notes

1. TMS consists in the application of a magnetic field on the scalp of the subject. The field pass through the skull and the meninx without being altered and turns into an electrical stimulation of the neuronal population underneath. Applied on the primary motor cortex the amplitude of the obtained motor evoked potentials reveals the state of excitability of the motor system.

2. These "co-speech gestures" are to be distinguished from the signs which form the elements of the signed languages employed by deaf communities. A sign is to be understand as a gesture that has been ritualized and hence has acquired a specific meaning within some community.

3. This phrasing is to emphasize that some apes have acquired a repertoire of hand signs but have not acquired the syntactic skills of assembling those signs in the fashion characteristic of a true human signed language.

References

Albright, Adam & Hayes Bruce (2003). Rules vs. analogy in English past tenses: A computational/experimental study. *Cognition*, *90*, 119–161.

Arbib, Michael A. (1981). Perceptual structures and distributed motor control. In Vernon B. Brooks (Ed.), *Handbook of physiology, Section 2: The nervous system, Vol. II, Motor control, Part 1* (pp. 1449–1480). American Physiological Society.

Arbib, Michael A. (2002). The mirror system, imitation, and the evolution of language. In Chrystopher Nehaniv & Kerstin Dautenhahn (Eds.), *Imitation in animals and artifacts* (pp. 229–280). Cambridge MA: The MIT Press.

Arbib, Michael A. (2003). Schema theory. In Michael A. Arbib (Ed.), *The handbook of brain theory and neural networks* (pp. 993–998) (Second Edition). Cambridge, MA: A Bradford Book/The MIT Press.

Arbib, Michael A. (2004). How far is language beyond our grasp? A response to Hurford. In D. Kimbrough Oller & Ulrike Griebel (Eds.), *Evolution of communication systems: A comparative approach* (pp. 315–321). Cambridge, MA: The MIT Press.

Arbib, Michael A. (2005a). From monkey-like action recognition to human language: An evolutionary framework for neurolinguistics. *Behavioral and Brain Sciences, 28*, 105–214.

Arbib, Michael A. (2005b). Interweaving protosign and protospeech: Further developments beyond the mirror. *Interaction Studies: Social Behaviour and Communication in Biological and Artificial Systems, 6(2)*, 145–171.

Arbib, Michael A. (2006). A sentence is to speech as what is to action? *Cortex, 42(2),* 507–514.
Arbib, Michael A., Aude Billard, Marco Iacoboni, & Erhan Oztop (2000). Synthetic brain imaging: Grasping, mirror neurons and imitation. *Neural Networks, 13,* 975–997.
Arbib, Michael A. & Mihail Bota (2003). Language evolution: Neural homologies and neuroinformatics. *Neural Networks, 16,* 1237–1260.
Arbib, Michael A. & Giacomo Rizzolatti (1997). Neural expectations: A possible evolutionary path from manual skills to language. *Communication and Cognition, 29,* 393–424.
Bates, Elizabeth & Frederic Dick (2002). Language, gesture, and the developing brain. *Developmental Psychobiology, 40,* 293–310.
Beck, Benjamin B. (1980). *Animal tool behavior: The use and manufacture of tools by animals.* New York: Garland.
Binkofski, Ferdinand, Giovanni Buccino, Stefan Posse, Rüdiger J. Seitz, Giacomo Rizzolatti, & Hans-Joachim Freund (1999a). A fronto-parietal circuit for object manipulation in man: Evidence from an fMRI study. *European Journal of Neuroscience, 11,* 3276–3286.
Binkofski, Ferdinand, Giovanni Buccino, Klaus M. Stephan, Giacomo Rizzolatti, Rüdiger J. Seitz, & Hans-Joachim Freund (1999b). A parieto-premotor network for object manipulation: Evidence from neuroimaging. *Experimental Brain Research, 128,* 21–31.
Bischoff-Grethe, Amanda, Michael G. Crowley, & Michael A. Arbib (2003). Movement inhibition and next sensory state prediction in basal ganglia. In Ann M. Graybiel, Mahlon R. Delong, & Stephen T. Kitai (Eds.), *The Basal Ganglia VI* (pp. 267–277). New York: Kluwer Academic/Plenum Publishers.
Boesch, Christophe & Hedwige Boesch (1983). Optimization of nut-cracking with natural hammers by wild chimpanzees. *Behavior,* 83, 265–286.
Bradshaw, John L. (1997). *Human evolution: A neuropsychological perspective.* Hove: Psychology Press.
Buccino, Giovanni, Ferdinand Binkofski, Gereon R. Fink, Luciano Fadiga, Vittorio Gallese, Rüdiger J. Seitz, Karl Zilles, Giacomo Rizzolatti, & Hans-Joachim Freund (2001). Action observation activates premotor and parietals areas in a somatotopic manner: A fMRI study. *European Journal of Neuroscience, 13,* 400–404.
Butterworth, George (2003). Pointing is the royal road to language for babies. In Sotaro Kita (Ed.), *Pointing: Where languages, culture and cognition meet* (pp. 9–34). Lawrence Erlbaum Associates.
Call, Josep & Michael Tomasello (1994). Production and comprehension of referential pointing by orangutans (*Pongo pygmaeus*). *Journal of Comparative Psychology, 108,* 307–317.
Campbell, Ruth, Mairead MacSweeney, Simon Surguladze, Gemma Calvert, Philip McGuire, John Suckling, Michael J. Brammer, & Anthony S. David (2001). Cortical substrates for the perception of face actions: An fMRI study of the specificity of activation for seen speech and for meaningless lower-face acts (gurning). *Cognitive Brain Research, 12,* 233–243.
Carr, Laurie, Marco Iacoboni, Marie-Charlotte Dubeau, John C. Mazziotta, & Gian-Luigi Lenzi (2003). Neural mechanisms of empathy in humans: A relay from neural systems for imitation to limbic areas. *Proceedings of the National Academy of Sciences, 100 (9),* 5497–5502.
Chaminade, Thierry, Andrew N. Meltzoff, & Jean Decety (2002). Does the end justify the means? A PET exploration of the mechanisms involved in human imitation. *Neuroimage, 15,* 318–328.
Corballis, Michael C. (2003). From mouth to hand: Gesture, speech, and the evolution of right-handedness. *Behavioral and Brain Sciences, 26,* 199–260.

Corina, David P., Howard Poizner, Ursula Bellugi, Todd Feinberg, Dorothy Dowd, & Lucinda O'Grady-Batch (1992). Dissociation between linguistic and nonlinguistic gestural systems: A case for compositionality. *Brain and Language, 43 (3)*, 414–447.

Deacon, Terrence (1997). *The symbolic species.* Allen Lane/The Penguin Press.

Decety, Jean & Thierry Chaminade (2003). Neural correlates of feeling sympathy. *Neuropsychologia, 41*, 127–138.

di Pellegrino, Giuseppe, Luciano Fadiga, Leonardo Fogassi, Vittorio Gallese, & Giacomo Rizzolatti (1992). Understanding motor events: A neurophysiological study. *Experimental Brain Research, 91 (2)*, 176–180.

Donald, Merlin (1999). Preconditions for the evolution of protolanguages. In Michael C. Corballis & Stephen E.G. Lea (Eds.), *The descent of mind* (pp. 138–154). Oxford: Oxford University Press.

Embick, David, Alec Marantz, Yasushi Miyashita, Wayne O'Neil, & Kuniyoshi L. Sakai (2000). A syntactic specialization for Broca's area. *Proceedings of the National Academy of Sciences of the United States of America, 97*, 6150–6154.

Fadiga, Luciano, Laila Craighero, Giovanni Buccino, & Giacomo Rizzolatti (2002). Speech listening specifically modulates the excitability of tongue muscles: A TMS study. *European Journal of Neuroscience, 15*, 399–402.

Fadiga, Luciano, Leonardo Fogassi, Giovanni Pavesi, & Giacomo Rizzolatti (1995). Motor facilitation during action observation: A magnetic stimulation study. *Journal of Neurophysiology, 73 (6)*, 2608–2611.

Falkenstein, Lorne. "Étienne Bonnot de Condillac». In Edward N. Zalta (Ed.), *The Stanford encyclopedia of philosophy* (Winter 2002 Edition). URL = <http://plato.stanford.edu/archives/win2002/entries/condillac/>.

Farnè, Alessandro, Alice C. Roy, Pascal Giraux, Jean-Michel Dubernard, & Angela Sirigu (2002). Face or hand, not both: Perceptual correlates of deafferentation in a former amputee. *Current Biology, 12 (15)*, 1342–1346.

Ferrari, Pier-Francesco, Vittorio Gallese, Giacomo Rizzolatti, & Leonardo Fogassi (2003). Mirror neurons responding to the observation of ingestive and communicative mouth actions in the monkey ventral premotor cortex. *European Journal of Neuroscience, 17*, 1703–1714.

Ferrari, Pier-Francesco, Stefano Rozzi, & Leonardo Fogassi (2005). Mirror neurons responding to observation of actions made with tools in monkey ventral premotor cortex. *Journal of Cognitive Neuroscience, 17 (2)*, 212–226.

Friederici, Angela D., Bertram Opitz, & D. Yves von Cramon (2000). Segregating semantic and syntactic aspects of processing in the human brain: An fMRI investigation of different words types. *Cerebral Cortex*, 10, 698–705.

Fujiyama, Asao, Hidemi Watanabe, Atsushi Toyoda, Todd D. Taylor, Takehiko Itoh, Shih-Feng Tsai, Hong-Seog Park, Marie-Laure Yaspo, Hans Lehrach, Zhu Chen, Gsng Fu, Naruya Saitou, Kazutoyo Osoegawa, Pieter J. de Jong, Yumiko Suto, Masahira Hattori, & Yoshiyuki Sakaki (2002). Construction and analysis of a human-chimpanzee comparative clone map. *Science, 295 (5552)*, 131–134.

Gallese, Vittorio, Luciano Fadiga, Leonardo Fogassi, & Giacomo Rizzolatti (1996). Action recognition in the premotor cortex. *Brain, 119 (2)*, 593–609.

Gallese, Vittorio, Luciano Fadiga, Leonardo Fogassi, & Giacomo Rizzolatti (2002). Action representation and the inferior pariatel lobule. In Wolfgang Prinz & Bernhard Hommel (Eds.),

Common mechanisms in perception and action: Attention and Performance, Vol. XIX (pp. 334–355). Oxford: Oxford University Press.

Gardner, R. Allen & Beatrix T. Gardner (1969). Teaching sign language to a chimpanzee. *Science, 165*, 664–672.

Gentilucci, Maurizio, Paola Santunione, Alice C. Roy, & Silvia Stefanini (2004). Execution and observation of bringing a fruit to the mouth affect syllable pronunciation. *European Journal of Neuroscience, 19*, 190–202.

Gentilucci, Maurizio, Silvia Stefanini, Alice C. Roy, & Paola Santunione (2004). Action observation and speech production: Study on children and adults. *Neuropsychologia, 42 (11)*, 1554–1567.

Gerardin, Emmanuel, Angela Sirigu, Stéphane Lehéricy, Jean-Baptiste Poline, Bertrand Gaymard, Claude Marsault, Yves Agid, & Denis Le Bihan (2000). Partially overlapping neural networks for real and imagined hand movements. *Cerebral Cortex, 10*, 1093–1104.

Ghazanfar, Asif A. & Marc D. Hauser (1999). The neuroethology of primate vocal communication: Substrates for the evolution of speech. *Trends in Cognitive Neurosciences, 3*, 377–384.

Giraux, Pascal, Angela Sirigu, Fabien Schneider, & Jean-Michel Dubernard (2001). Cortical reorganization in motor cortex after graft of both hands. *Nature Neuroscience, 4*, 1–2.

Goldenberg, Georg & Sonja Hagmann (1998). Tool use and mechanical problem solving in apraxia. *Neuropsychologia, 36 (7)*, 581–89.

Goldin-Meadow, Susan (2002). Constructing communication by hand. *Cognitive Development, 17*, 1385–1405.

Goldin-Meadow, Susan (1999). The role of gesture in communication and thinking. *Trends in Cognitive Sciences, 3 (11)*, 419–429.

Goldin-Meadow, Susan & Cynthia Butcher (2003). Pointing toward two-word speech in young children. In Sotaro Kita (Ed), *Pointing: Where languages, culture and cognition meet* (pp. 85–108). Lawrence Erlbaum associates.

Grafton, Scott T., Michael A. Arbib, Luciano Fadiga, & Giacomo Rizzolatti (1996). Localization of grasp representations in humans by PET: 2. Observation compared with imagination. *Experimental Brain Research, 112*, 103–111.

Grèzes, Julie, Jorge L. Armony, James Rowe, & Richard E. Passingham (2003). Activations related to "mirror" and "canonical" neurones in the human brain: An fMRI study. *Neuroimage, 18*, 928–937.

Grèzes, Julie, Nicolas Costes, & Jean Decety (1999). The effects of learning and intention on the neural network involved in the perception of meaningless actions. *Brain, 122*, 1875–1887.

Gruber, Oliver, Peter Indefrey, Helmuth Steinmetz, & Andreas Kleinschmidt (2001). Dissociating neural correlates of cognitive components in mental calculation. *Cerebral Cortex, 11*, 350–359.

Hamzei, Farsin, Michel Rijntjes, Christian Dettmers, Volkmar Glauche, Cornelius Weiller, & Christian Buchel (2003). The human action recognition system and its relationship to Broca's area: An fMRI study. *Neuroimage, 19*, 637–644.

Harrington, Deborah L. & Kathleen Y. Haaland (1992). Motor sequencing with left hemisphere damage: Are some cognitive deficits specific to limb apraxia? *Brain*, 115, 857–874.

Hashimoto, Ryuichiro & Kuniyoshi L. Sakai (2002). Specialization in the left prefrontal cortex for sentence comprehension. *Neuron, 35*, 589–597.

Hauser, Marc D., Noam Chomsky, & W. Tecumseh Fitch (2002). The faculty of language: What is it, who has it, and how did it evolve? *Science, 298*, 1569–1579.

Hayes, Cathy (1951). *The ape in our house*. New York: Harper.
Heim, Steven, Bertram Opitz, & Angela D. Friederici (2003). Distributed cortical networks for syntax processing: Broca's area as the common denominator. *Brain and Language, 85*, 402–408.
Heiser, Marc, Marco Iacoboni, Fumiko Maeda, Jake Marcus, & John C. Mazziotta (2003). The essential role of Broca's area in imitation. *European Journal of Neuroscience, 17*, 1123–1128.
Hihara, Sayaka, Shigeru Obayashi, Michio Tanaka, & Atsushi Iriki (2003). Rapid learning of sequential tool use by macaque monkeys. *Physiology and Behavior, 78*, 427–434.
Hihara, Sayaka, Hiroko Yamada, Atsushi Iriki, & Kazuo Okanoya (2003). Spontaneous vocal differentiation of coo-calls for tools and food in Japanese monkeys. *Neuroscience Research, 45*, 383–389.
Hopkins, William D. & David A. Leavens (1998). Hand use and gestural communication in chimpanzees (*pan troglodytes*). *Journal of Comparative Psychology, 112 (1)*, 95–99.
Hurford, James (2004). Language beyond our grasp: What mirror neurons can, and cannot, do for language evolution. In D. Kimbrough Oller & Ulrike Griebel (Eds.), *Evolution of communication systems: A comparative approach* (pp. 297–313). Cambridge, MA: The MIT Press.
Iacoboni, Marco, Roger Woods, Marcel Brass, Harold Bekkering, John C. Mazziotta, & Giacomo Rizzolatti (1999). Cortical mechanisms of human imitation. *Science, 286*, 2526–2528.
Iriki, Atsushi, Michio Tanaka, & Yoshiaki Iwamura (1996). Coding of modified body schema during tool use by macaque postcentral neurones. *Neuroreport, 7*, 2325–2330.
Iverson, Jana M. & Susan Goldin-Meadow (1998). Why people gesture when they speak. *Nature, 396*, 228.
Johnson-Frey, Scott H. (2004). The neural bases of complex tool use in humans. *Trends in Cognitive Sciences, 8*, 71–78.
Johnson-Frey, Scott H. (2003). What's so special about human tool use? *Neuron, 39*, 201–204.
Johnson-Frey, Scott H., Farah R. Maloof, Roger Newman-Norlund, Chloe Farrer, Souheil Inati, & Scott T. Grafton (2003). Actions or hand-object interactions? Human inferior frontal cortex and action observation. *Neuron, 39*, 1053–1058.
Kang, A. Min, R. Todd Constable, John C. Gore, & Sergey Avrutin (1999). An event-related fMRI study on implicit phrase-level syntactic and semantic processing. *Neuroimage, 10*, 555–561.
Kellogg, Winthrop N. & Luella A. Kellogg (1933) *The ape and the child*. New York: McGraw Hill.
King, Barbara J. (1996). Syntax and language origins. *Language and Communication, 16*, 193–203.
Knecht, Stefan, Michael Deppe, Bianca Drager, Lars Bobe, Hubertus Lohmann, E. Bernd Ringelstein, & Henning Henningsen (2000). Language lateralization in healthy right-handers. *Brain, 123*, 74–81.
Kohler, Evelyne, Christian Keysers, M. Alessandra Umiltà, Leonardo Fogassi, Vittorio Gallese, & Giacomo Rizzolatti (2002). Hearing sounds, understanding actions: Action representation in mirror neurons. *Science, 297*, 846–848.
Koski, Lisa, Marco Iacoboni, Marie-Charlotte Dubeau, Roger P. Woods, & Mazziotta John C. (2003). Modulation of cortical activity during different imitative behaviors. *Journal of Neurophysiology, 89*, 460–471.

Koski, Lisa, Andreas Wohlschlager, Harold Bekkering, Roger P. Woods, Marie-Charlotte Dubeau, John C. Mazziotta, & Marco Iacoboni (2002). Modulation of motor and premotor activity during imitation of target-directed actions. *Cerebral Cortex, 12*, 847–855.

Lacquaniti, Francesco, Daniela Perani, Emmanuel Guignon, Valentino Bettinardi, Marco Carrozzo, F. Grassi, Yves Rossetti, & Ferruccio Fazio (1997). Visuomotor transformations for reaching to memorized targets: A PET study. *NeuroImage, 5*, 129–146.

Leavens, David A. (2003). Integration of visual and vocal communication: Evidence for miocene origins. *Behavioral Brain Sciences, 26*, 232.

Leavens, David A. & Hopkins William D. (1998). Intentional communication by chimpanzees: A cross-sectional study of the use of referential gestures. *Developmental Psychology, 34*, 813–822.

Leavens, David A. & Hopkins William D. (1999). The whole-hand point: The structure and function of pointing from a comparative perspective. *Journal of Comparative Psychology, 113*, 417–425.

Leiguarda, Ramon C. & C. David Marsden (2000). Limb apraxias: Higher-order disorders of sensorimotor integration. *Brain, 123*, 860–879.

Leslie, Kenneth R., Scott H. Johnson-Frey, & Scott T. Grafton (2004). Functional imaging of face and hand imitation: Towards a motor theory of empathy. *Neuroimage, 21*, 601–607.

Liberman, Alvin M. & Doug H. Whalen (2000). On the relation of speech to language. *Trends in Cognitive Sciences, 4*, 187–196.

Liberman, Alvin M. & Ignatius G. Mattingly (1985). The motor theory of speech perception revisited. *Cognition, 21*, 1–36.

Liebal, Katja, Simone Pika, & Michael Tomasello (2004). Social communication in siamangs (*Symphalangus syndactylus*): Use of gestures and facial expressions. *Primates, 45*, 41–57.

Liepmann, Hugo (1913). Motor aphasia, anarthria and apraxia. *Proceedings of the 17th International Congress of Medicine, Part 2* (pp. 97–106). London.

Maess, Burkhard, Stefan Koelsch, Thomas C. Gunter, & Angela. D. Friederici (2001). Musical syntax is processed in Broca's area: An MEG study. *Nature Neuroscience, 4*, 540–545.

Marshall, Jane, Jo Atkinson, Elaine Smulovitch, Alice Thacker, & Bencie Woll (2004). Aphasia in a user of British Sign Language: Dissociation between sign and gesture. *Cognitive Neuropsychology, 21*, 537–554.

Matsumura, Michikazu, R. Kawashima, Eiichi Naito, K. Satoh, T. Takahashi, T. Yanagisawa, & H. Fukuda (1996). Changes in rCBF during grasping in humans examined by PET. *NeuroReport, 7*, 749–752.

McGrew, W. C. & L. F. Marchant (2001). Ethiological study of manual laterality in the chimpanzees of the Mahale mountains, Tanzania. *Behaviour, 138 (3)*, 329–358.

McNeill, David (1992). *Hand and mind: What gestures reveal about thought*. Chicago: University of Chicago Press.

Meador, Kimford J., David W. Loring, K. Lee, M. Hughes, G. Lee, M. Nichols, & Kenneth M. Heilman (1999). Cerebral lateralization: Relationship of language and ideomotor praxis. *Neurology, 53*, 2028–2031.

Mehler, M. F. (1987). Visuo-imitative apraxia. *Neurology, 37*, 129.

Moro, Andrea, Marco Tettamanti, Daniela Perani, C. Donati, Stefano Cappa, & Fazio Ferruccio (2001). Syntax and the brain: Disentangling grammar by selective anomalies. *Neuroimage, 13*, 110–118.

Newman, Sharlene D., Marcel A. Just, Timothy A. Keller, Jennifer Roth, & Patricia A. Carpenter (2003). Differential effects of syntactic and semantic processing on the subregions of Broca's area. *Cognitive Brain Research, 16*, 297–307.

Nishimura, Takeshi, Akichika Mikami, Juri Suzuki, & Tetsuro Matsuzawa (2003). Descent of the larynx in chimpanzee infants. *Proceedings of the National Academy of Sciences, 100*, 6930–6933.

Nishitani, Nobuyuki & Riitta Hari (2000). Temporal dynamics of cortical representation for action. *Proceedings of the National Academy of Sciences, 97*, 913–918.

Palmer, A. R. (2002). Chimpanzee right-handedness reconsidered: Evaluating the evidence with funnel plots. *American Journal of Physical Anthropology, 121*, 382–384.

Papagno, Constanza, Della Sala Sergio, & Basso Anna (1993). Ideomotor apraxia without aphasia and aphasia without apraxia: The anatomical support for a double dissociation. *Journal of Neurology, Neurosurgery and Psychiatry, 56*, 286–289.

Passingham, Richard (1993). *The frontal lobes and voluntary action.* Oxford: Oxford University Press.

Pika, Simone, Katja Liebal, & Michael Tomasello (2003). Gestural communication in young gorillas (*Gorilla gorilla*): Gestural repertoire, learning, and use. *American Journal of Primatology, 60*, 95–111.

Pika, Simone, Katja Liebal, & Michael Tomasello (2005). Gestural communication in subadult bonobos (*Pan paniscus*): Repertoire and use. *American Journal of Primatology, 65*, 39–61.

Pinker, Steven (1997). Language as a psychological adaptation. *Ciba Foundation Symposium, 208*, 162–172.

Poizner, Howard, Alma S. Merians, Maryann A. Clark, Beth Macauley, Leslie J.G. Rothi, & Kenneth M. Heilman (1998). Left hemispheric specialization for learned, skilled, and purposeful action. *Neuropsychology, 12*, 163–182.

Povinelli, Daniel J., Laura A. Theall, James E. Reaux, & Sarah Dunphy-Lelii (2003). Chimpanzees spontaneously alter the location of their gestures to match the attentional orientation of others. *Animal Behaviour, 65*, 1–9.

Rapcsak, Steven Z., Cynthia Ochipa, Kathleen C. Anderson, & Howard Poizner (1995). Progressive ideomotor apraxia: Evidence for a selective impairment of the action production system. *Brain and Cognition, 27*, 213–236.

Rizzolatti, Giacomo & Michael A. Arbib (1998). Language within our grasp. *Trends in Neurosciences, 21*, 188–194.

Rizzolatti, Giacomo & Giovanni Buccino (2004). The mirror-neuron system and its role in imitation and language. In Stanislas Dehaene, Jean-Rene Duhamel, Marc Hauser, & Giacomo Rizzolatti (Eds.), *From monkey brain to human brain.* Cambridge, Massachusetts: MIT Press.

Rizzolatti, Giacomo, Luciano Fadiga, Vittotio Gallese, & Leonardo Fogassi (1996). Premotor cortex and the recognition of motor actions. *Cognitive Brain Research, 3*, 131–141.

Rizzolatti, Giacomo, Leonardo Fogassi, & Vittorio Gallese (2001). Neurophysiological mechanisms underlying the understanding and imitation of action. *Nature Reviews Neuroscience, 2*, 661–670.

Roth, R. R. (1995). *A study during sexual behavior in bonobo (Pan paniscus).* Calgary, University of Calgary Press.

Rumiati, Raffaella I., Peter H. Weiss, Tim Shallice, Giovanni Ottoboni, Johannes Noth, Karl Zilles, & Gereon R. Fink (2004). Neural basis of pantomiming the use of visually presented objects. *Neuroimage, 21*, 1224–1231.

Sandler, Wendy, Irit Meir, Carol Padden, & Mark Aronoff (2005). The emergence of grammar: Systematic structure in a new language. *PNAS, 102*, 2661–2665.

Savage-Rumbaugh, E. Sue, Stuart C. Shanker, & Talbot J. Taylor (1998). *Apes, language and the human mind*. Oxford: Oxford University Press.

Savage-Rumbaugh, E. Sue, B.J. Wilkerson, & R. Bakeman (1977). Spontaneous gestural communication among conspecifics in the pigmy chimpanzee (*Pan paniscus*). In Geoffrey H. Bourne (Ed.), *Progress in ape research* (pp. 97–116). New York: Academic Press.

Schnider, Armin, Robert E. Hanlon, David N. Alexander, & D. Frank Benson (1997). Ideomotor apraxia: Behavioral dimensions and neuroanatomical basis. *Brain and Language, 58*, 125–136.

Seidenberg, Mark S. (1997). Language acquisition and use: Learning and applying probabilistic constraints. *Science, 275*, 1599–1603.

Senghas, Ann, Sotaro Kita, & Asli Özyürek (2004). Children creating core properties of language: Evidence from an emerging sign language in Nicaragua. *Science, 305*, 1779–1782.

Studdert-Kennedy, Michael (2002). Mirror neurons, vocal imitation, and the evolution of particulate speech. In Maxim I. Stamenov & Vittorio Gallese (Eds.), *Mirror neurons and the evolution of brain and language* (pp. 207–227). Amsterdam: John Benjamins.

Szaflarski, Jerzy P., Jeffrey R. Binder, Edward T. Possing, Kristen A. McKiernan, B. Douglas Ward, & Thomas A. Hammeke (2002). Language lateralization in left-handed and ambidextrous people: fMRI data. *Neurology, 59*, 238–244.

Tanaka, Shigeki & Toshio Inui (2002). Cortical involvement for action imitation of hand/arm postures versus finger configurations: An fMRI study. *NeuroReport*, 13, 1599–1602.

Tanner, Joanne E. & Richard Byrne (1996). Representation of action through iconic gesture in a captive lowland gorilla. *Current Anthropology, 37*, 162–173.

Tomasello, Michael (1999). The human adaptation for culture. *Annual Review of Anthropology, 28*, 509–529.

Tomasello, Michael (2006). Why don't apes point? In N.J. Enfield & S.C. Levinson (Eds). *Roots of Human Sociality: Culture, Cognition and Interaction*, pp. 506–524. Oxford: Berg.

Tomasello, Michael & Josep Call (2004). The role of humans in the cognitive development of apes revisited. *Animal Cognition, 7*, 213–215.

Tomasello, Michael & Klaus Zuberbühler (2002). Primate vocal and gestural communication. In Gordon M. Burghardt (Ed.), *The cognitive animal: Empirical and theoretical perspectives on animal cognition* (pp. 293–299). Cambridge: MIT Press.

Umiltá, M. Alessandra, Evelyne Kohler, Vittorio Gallese, Leonardo Fogassi, Luciano Fadiga, Christian Keysers, & Giacomo Rizzolatti (2001). I know what you are doing: A neurophysiological study. *Neuron, 31*, 155–165.

Vea, Joachim & Jordi Sabater-Pi (1998). Spontaneous pointing behaviour in the wild pygmy chimpanzee (*Pan paniscus*). *Folia Primatologica, 69*, 289–290.

Waters, Gabriel S. & Roger S. Fouts (2002). Sympathetic mouth movements accompanying fine motor movements in chimpanzees (*Pan troglodytes*) with implications toward the evolution of language. *Neurological Research, 24*, 174–180.

Part II

Gestural communication in non-human primates

The gestural communication of apes

Simone Pika[1,3], Katja Liebal[2,3], Josep Call[3], and Michael Tomasello[3]
[1]University of Machester / [2]University of Portsmouth / [3]Max Planck Institute for Evolutionary Anthropology, Leipzig

Gestural communication of primates may allow insight into the evolutionary scenario of human communication given the flexible use and learning of gestures as opposed to vocalizations. This paper provides an overview of the work on the gestural communication of apes with the focus on their repertoire, learning mechanisms, and the flexibility of gesture use during interactions with conspecifics. Although there is a variation between the species in the types and numbers of gestures performed, the influence of ecology, social structure and cognitive skills on their gestural repertoires is relatively restricted. As opposed to humans, ape's gestures do not show the symbolic or conventionalized features of human gestural communication. However, since the gestural repertoires of apes are characterized by a high degree of individual variability and flexibility of use as opposed to their vocalizations it seems plausible that gestures were the modality within which symbolic communication first evolved.

Human communication is unique in the animal kingdom in any number of ways. Most importantly, of course, human communication depends crucially on linguistic symbols, which, to our knowledge, are not used by any other species in their natural environment. Although there is no universally agreed upon definition of linguistic symbols, many theorists would agree that they are, in their essence, individually learned and intersubjectively shared social conventions used to direct the attentional and mental states of others to outside entities referentially.

In looking for the evolutionary roots of human language, researchers quite naturally looked first at primate vocalizations. The groundbreaking discovery that vervet monkeys use different alarm calls in association with different predators (leading to different escape responses in receivers) raised the possibility that some nonhuman species may, like humans, use vocalizations to make reference to outside entities (Cheney & Seyfarth, 1990). But it has turned out since then that alarm calls of this type have arisen numerous times in evolution in species that also must organize different escape responses for different predators, including most prominently prairie dogs and domestic chickens (see Owings & Morton, 1998, for a review). It is also the case that primate vocalizations in general are unlearned and show very little flexibility of use: infants reared in social isolation still produce basically all of their species-typical call types from soon after birth (see Snowdon et al., 1997, for a review), and rearing individuals within the social context of another primate species produces no significant changes in the vocal repertoire (Owren et al., 1992). And importantly, there is currently no evidence that any species of ape has such referent specific alarm calls or any other vocalizations that appear to be referential (Cheney & Wrangham, 1987; see Crockford & Boesch, 2003, for context-specific calls) — which means that it is highly unlikely that vervet monkey alarm calls could be the direct precursor of human language — unless at some point apes used similar calls and have now lost them (however, see Slocombe & Zuberbühler, 2005).

But human communication is also unique in the way it employs manual and other bodily gestures. For example, to our knowledge only human beings gesture triadically, (that is for persons to external entities — the basic form of gestural reference), simply to share attention or comment on things.[1] And humans use other kinds of symbolic gestures as well, ranging from waving goodbye to signaling "OK" to conventionalized obscenities — which, to our knowledge, are also unique to the species. In general, one might say that human gestures are used functionally in ways very similar to language (e.g., symbolically, referentially, based on intersubjectively learned and shared social conventions) and many of the aspects of human linguistic communication that make it so different from primate vocalizations are also present in human gestures. The question thus arises: what is the nature of the gestural communication of primates, and how do they relate to human gestures and language? This question has received surprisingly little research attention, that is, outside our own research group and a few others.

Our research group has been studying the gestural communication of primates for about two decades. We have focused on their natural communication with one another, not with their communication with humans (for interesting work of this type see e.g., Gomez, 1990; Leavens & Hopkins, 1998). The vast majority of our earlier work focused on chimpanzees (*Pan troglodytes*), one of humans' two closest primate relatives, but more recently we have expanded our work to cover other ape species. In the current paper, we provide a summary of that work — beginning with primate gestural communication in general, based mainly on our extensive work with chimpanzees. We then briefly summarize our more recent work with other ape species. In all of this we focus especially on those aspects that might be of greatest interest to researchers investigating human gestural communication.

Primate gestural communication

Primates communicate using manual and bodily gestures mainly in relative intimate social contexts such as play, grooming, nursing, and during sexual and agonistic encounters. These are in general less evolutionarily urgent functions than those signaled by acts of vocal communication (e.g., avoiding predators, defending against aggressors, traveling as a group, discovering food), and perhaps as a result primates tend to use their gestures more flexibly than their vocalizations (Tomasello & Zuberbühler, 2002). Thus, unlike the case of vocal signals, there is good evidence that many primate gestures, especially those of the great apes, are individually learned and used quite flexibly. The individuals of some ape species may even on occasion invent new gestural signals (Goodall, 1986; Tomasello et al., 1985; Pika et al., 2003), and apes raised by humans sometimes learn some human-like gestures (Tomasello & Camaioni, 1997). However, the gestural communication of primates still shows few signs of referentiality (however, see Plooij, 1987; Pika & Mitani, 2006) or symbolicity, and so the questions arise: What is the nature of primate gestures? How are they learned and used?

Our work over the last 20 years has focused mainly on chimpanzees. Based on a number of lines of evidence, both naturalistic and experimental, it seems clear that chimpanzees most often learn their gestural signals not via imitation but rather via an individual learning process called 'ontogenetic ritualization' (Tomasello, 1996). In ontogenetic ritualization two organisms essentially shape one another's behavior in repeated instances of a social interaction. The general form of this type of learning is:

- Individual A performs behavior X;
- Individual B reacts consistently with behavior Y;
- Subsequently B anticipates A's performance of X, on the basis of its initial step, by performing Y; and
- Subsequently, A anticipates B's anticipation and produces the initial step in a ritualized form (waiting for a response) *in order to* elicit Y.

For example, *play hitting* is an important part of the rough-and-tumble play of chimpanzees, and so many individuals come to use a stylized *arm raise* to indicate that they are about to hit the other and thus initiate play (Goodall, 1986). An example from human infants is when they raise their arms to be picked up, which is not learned by imitating other infants but rather is ritualized from the picking up process itself (Lock, 1978). The main point in ritualization is that a behavior that was not at first a communicative signal becomes one by virtue of the anticipations of the interactants over time. There is no evidence that any primate species acquires the majority of its gestural signals by means of imitative learning (Tomasello & Call, 1997), which is normally required for the forming of a true communicative convention — although there may be some exceptions in the case of individual gestures (see Nishida, 1980; McGrew & Tutin, 1978 for group-specific gestures of chimpanzees in the wild).

In addition, we have also investigated whether chimpanzees, like human infants, use their gestures "intentionally" and flexibly (Piaget, 1952; Bates, 1976; Bruner, 1981). The criterion most often used with human infants concern means-ends dissociation, characterized by the flexible relation of signaling behavior to the recipient and goal, for example, an individual uses a single gesture for several goals (*touch* for nursing and riding) or different gestures for the same goal (*slap ground* and *bodybeat* for play). With regard to such flexibility of use, Tomasello et al. (1994, 1997) found that many chimpanzee gestures were used in multiple contexts, sometimes across widely divergent behavioral domains. Also, sometimes different gestures were used in the same context interchangeably toward the same end — and individuals sometimes performed these in rapid succession in the same context (e.g., initiating play first with a *poke-at* followed by an *arm-raise*). In some instances both monkeys and apes have been observed to use some gestures in a way that suggests 'tactical deception', which indicates that a gesture was used outside its ordinary context (Whiten & Byrne, 1988).

Another important issue concerning flexibility of use is so-called audience effects, that is, differential use of gestures or other communicative signals as a function of the psychological states of the recipient. Tomasello et al. (1994, 1997) found that chimpanzee juveniles only give a visual signal to solicit play (e.g., *arm-raise*) when the recipient is already oriented appropriately, but they use their most insistent attention-getter, a physical *poke-at*, most often when the recipient is socially engaged with others. Tanner and Byrne (1993) reported that a female gorilla repeatedly used her hands to hide her playface from a potential partner, indicating some flexible control of the otherwise involuntary grimace — as well as a possible understanding of the role of visual attention in the process of gestural communication. Furthermore, in an experimental setting, Call and Tomasello (1994) found that some orangutans also were sensitive to the gaze direction of their communicative partner, choosing not to communicate when the partner was not oriented to them. In addition, Kummer (1968) reported that before they set off foraging, male hamadryas baboons engage in "notifying behavior" in which they approach another individual and look directly into her. Presumably, they use this behavior to make sure that the other is looking before the trek begins. Overall, audience effects are very clear in primate gestural communication, but these all concern whether others can or cannot see the gesture — i.e., are bodily oriented toward the gesturer — not the particular knowledge states of others (as is common in human communication).

Chimpanzees employ basically two types of intentional gesture. First are "incipient actions" that have become ritualized into gestures (see Tinbergen, 1951, on "intention-movements"). For example, as noted above, many juveniles come to use a stylized *arm-raise* to initiate play, ritualized from actual acts of play hitting in the context of rough-and-tumble play. Many youngsters also ritualize signals for asking their mother to lower her back so they can climb on, for example, a brief *touch* on the top of the rear end, ritualized from occasions on which they pushed her rear end down mechanically. Infants often do something similar, such as a light *touch* on the arm (ritualized from actually pulling the arm), to ask their mothers to move it so as to allow nursing. Interestingly, Tanner and Byrne (1996) described a number of gestures in gorillas that they interpret as iconic (depict motion in space). That is, an adult male gorilla often seemed to indicate to a female playmate iconically, using his arms or whole body, the direction in which he wanted her to move, the location he wanted her to go to, or the action he wanted her to perform. However, these might simply be normal ritualized incipient actions with the iconicity being in the eyes of the human only; in fact, a role for iconicity in gorillas' and other apes' comprehension of gestures has not at this point been demonstrated (Tomasello & Call, 1997; Pika et al., 2003).

The second type of intentional gestures are "attractors" (or attention-getters) aimed at getting others to look at the self. For example, a well-known behavior from the wild is the *leaf-clipping* of adult males, which serves to make a noise that attracts the attention of females to their sexual arousal (Nishida, 1980). Similarly, when youngsters want to initiate play they often attract the attention of a partner to themselves by *slapping the ground* in front of, *poking at*, or *throwing* things at the desired partner (Tomasello, Gust, & Frost, 1989). Because their function is limited to attracting the attention of others, attractors most often attain their specific communicative goal from their combination with seemingly involuntary displays. That is, the specific desire to play or mate is communicated by the 'play-face' or penile erection, with the attractor serving only to gain attention to it.

On the surface, attractors would seem to bear some relation to dyadic deictic gestures that simply point out things in the environment, and incipient actions would seem at least somewhat similar to lexical symbols that have relatively context-dependent semantic content. But the primate versions are obviously different from the human versions as well, most especially because the primate versions are dyadic and not referential. Attractors are thus really most similar not to deictics, which are referential, but to human attention-getters like "Hey!" that simply serve to make sure that a communicative channel is open, or else emphasizes a gesture. Incipient actions are most similar to certain kinds of ritualized performatives — for example, greetings and some imperatives — that serve to regulate social interactions, not refer to or comment upon anything external. It is also interesting that systematic observations of chimpanzee gesture combinations reveal no evidence of a strategy in which they first use an attractor to make sure the other is looking followed by an incipient action containing specific semantic content (vaguely analogous to topic-comment structure; Liebal, Call, & Tomasello, 2004). One would think that if chimpanzees understood the different communicative functions of these two types of gesture, this kind of combination would be relatively frequent.[2]

Importantly in the current context, virtually all of the intentional gestures used by chimpanzees share two important characteristics that make them crucially different from human deictic and symbolic gestures. First of all, they are almost invariably used in dyadic contexts (the one major exception is noted below). That is, attractors are used to attract the attention of others to the self, not triadically, to attract the attention of others to some outside entity. Likewise, incipient-movement gestures are used dyadically to request some behavior of another toward the self (e.g., play, grooming, sex), not to request behavior directed to some entity in the external environment. This almost exclusive dyadic use is different from the behavior of human infants who gesture triadically from their very first attempts in addition to dyadic gestures (Carpenter, Nagell, & Tomasello, 1998).

Second and relatedly, chimpanzee gestures, both attractors and incipient-movements, seem to be used exclusively for imperative purposes, that means to request actions from others. They do not seem to be used declaratively to direct the attention of others to an outside object or event, simply for the sake of sharing interest in it or commenting on it. Most clearly, chimpanzees in their natural habitats have not been observed to draw attention to objects in the typically human ways of pointing to them or holding them up for showing to others (Tomasello & Call, 1994). However, Pika and Mitani (2006) observed the widespread use of a gesture in male chimpanzees in the wild, the *directed scratch*. This gesture seems to be used to indicate a precise spot on the body to be groomed, and may qualify as referential. According to Menzel (1973), "one good reason that chimpanzees very seldom point manually is that they do no have to": being quadrupedal, their whole body is pointing (Plooij, 1987). Human infants, however, produce gestures for both imperative and declarative purposes from early in their communicative development.

Overall, the almost exclusive use of dyadic gestures for imperative purposes is consistent with the view that chimpanzees mostly do not use their gestures symbolically, that is, in intersubjective acts of reference.

The one major exception to this pattern of chimpanzee gestures as dyadic and imperative (and mainly produced in close physical proximity) is *food-begging*, in which youngsters attempt to obtain food from adults.[3] Infants beg for food by a number of related means, some of which do not involve communicative signals, such as: directly grabbing the food, staring at the food or into the eyes of the adult from very close range, sucking on the lower lip of the adult, rubbing the adult's chin as she is chewing the food, and so forth. In addition, however, infants sometimes hold out their hand, palm up, under the mother's chin (see Bard, 1992, for a similar behavior in infant orangutans). This *palm-up* gesture is clearly triadic — it is a request to another for food — and it is somewhat distal since the signaler is not touching the recipient. It should be noted, however, that food begging happens in very close physical proximity, with much touching, and that the *palm-up* gesture is likely ritualized from the rubbing of the chin. And it is still an imperative gesture, of course, since the goal of obtaining food is clear. Nevertheless, this *food-begging* gesture demonstrates that in some circumstances chimpanzees can ritualize some triadic and moderately distal gestures for purposes of obtaining things from others.

Overall, chimpanzee and other primate gestural communication clearly shows more flexibility of use than primate vocal communication, perhaps because it concerns less evolutionarily urgent activities than those associated with vocalizations. Apes in particular create new gestures routinely, and in general use many of their gestures for multiple communicative ends. Audience effects are also integral to ape gestural communication and concern more than simple presence-absence of others — but only in the sense of whether others are in a position to see the gesture. Overall, then, we have much evidence that primates use their gestures much more flexibly than their vocal signals. But we still have very little evidence that they use their gestures symbolically.

A comparison of apes

Most of the general description just given was based on work with chimpanzees, with only a minority of observations from other primate species. Recently our research group has focused systematically on the gestural communication of the other three great ape species, along with one species of small ape, respectively: bonobos (*Pan paniscus;* Pika & Tomasello, 2005), gorillas (*Gorilla gorilla;* Pika et al., 2003), orangutans (*Pongo pygmaeus;* Liebal et al., 2006), and siamangs (*Symphalangus syndactulus* — one of approximately a dozen species of gibbon; Liebal et al., 2004). For current purposes, our main question is whether the chimpanzee pattern is also characteristic of these species.

This is certainly not a foregone conclusion as there have been a number of proposals to the effect that the nature of the communication of different species should be a function of (1) the ecology of the species, (2) the social structure of the species, and (3) the cognitive skills of the species. These apes vary from one another greatly in all of these dimensions. For example, in terms of ecology it has been proposed that vocal communication predominates in arboreal species, when visual access to conspecifics is poor, whereas gestural communication predominates in more terrestrial species (Marler, 1965). In the apes, the orangutans and siamangs are almost totally arboreal, bonobos and chimpanzees divide their time between the ground and the trees, and gorillas are mainly terrestrial. In terms of social structure, it has been proposed that species with a more despotic social structure in which the outcome of most social interactions is, in a sense, predetermined should have a smaller repertoire of gestural signals, whereas species with a more egalitarian social structure involving more complex and negotiated social interactions should have a larger repertoire of gestural signals (Maestripieri, 1997). In the apes, gorillas perhaps tend toward the more despotic, whereas bonobos are more egalitarian. In terms of cognitive skills, we really do not have enough information to know if apes differ from one another in ways relevant for communication.

The methods of observation and analysis used in our studies derive ultimately from the series of studies on chimpanzee gestural communication conducted by Tomasello and colleagues over a dozen year period (Tomasello et al., 1985, 1989, 1994, 1997). We also conducted a follow-up study focused on the issue of gesture combinations (Liebal et al., 2004). The precise methods used evolved during this time period, and so the methods used in the recent studies are based most directly on the two studies from the 1990's and the follow-up study. Of special importance, only the follow-up study used focal animal sampling — observers watch a particular individual for a specific length of time no matter what it is doing — and so only it can be used to estimate absolute frequencies (the earlier studies used scan sampling in which observers simply looked for occurrences of target behaviors from anyone in the group). All of the studies summarized here used either focal animal sampling, or some combination of focal animal and behavior sampling (see Call & Tomasello, 2007, for details). To count in our observations, we had to observe an individual produce the same gesture on more than one occasion. In all five species, individuals from several different captive groups were observed.

Most of our observations and analyses have focused on three major issues. First is the goal-directed or intentional nature of particular gestures, operationalized as flexibility of use. We thus want to know such things as the variability in the gestural repertoires of different individuals, as an indication of the degree to which there is a fixed set in the species. Perhaps of most direct relevance to issues of flexibility, we want to identify gestures that are used by the same individual in multiple behavioral contexts, and also to identify contexts in which the same individual uses multiple gestures.

The second issue is how particular gestures are learned. In the absence of experimental interventions, we will again be interested in individual differences as an indication of whether gestures are learned or not learned — or perhaps even invented, as signals used by only one individual would seem to indicate individual invention. But most directly, we are concerned with whether particular gestures are ontogenetically ritualized in something like the manner outlined above, or whether, alternatively, they are socially learned from others using one or another form of imitation. In general, signals used by all or most members of one group, but not by the members of any other group of the same species, would seem to suggest some of type of social learning or imitation. Conversely, if the variability in individual gestural repertoires within a group is just as large as that between groups of the same species, then it is very unlikely that social learning or imitation is the major learning process — and much more likely that ontogenetic ritualization is what has occurred.

The third issue is adjustments for audience. As noted above, it is fairly common for primate species to produce particular gestural signals only when certain types of individuals are present — and indeed such audience effects are also characteristic of the vocal signaling of some primal species as well (e.g., domestic chickens; Evans, Evans, & Marler, 1993). But our more specific concern is with the question of whether an individual chooses a particular signal category depending on the attentional state of a particular recipient. For example, we are interested in whether individuals use visual gestures only when the potential recipient is visually oriented to them, and whether they use tactile signals preferentially when the potential recipient is not visually oriented to them. Such adjustments would seem to indicate that the signaler knows something about how its signal is being perceived by the recipient.

Repertoire and use

Perhaps the most basic comparative question is the relative sizes of the gesture repertoire of the different species. Our two nearest ape relatives, chimpanzees and bonobos, display between 20 and 30 gesture types across all groups studied, with particular individuals using, on average, about 10 gestures each from the species-wide pool. This pattern also holds for siamangs and indicates relative high individual variability. Gorillas and orangutans are at the high end of this repertoire size across groups (± 30), but individuals in the species are more similar to one another as their individual repertoire sizes are close to 20, roughly double the size of the two *Pan* species.

In terms of flexibility of use, we may look first from the perspective of functional contexts such as play, nursing, travelling, etc. Chimpanzees, bonobos, gorillas, and siamangs use an average of two to three gestures per functional context. Orangutans, on the other hand, use about five different gestures per functional context. Looking from the opposite perspective, we can ask in how many contexts each gesture is used. In this case, chimpanzees, orangutans, and siamangs, used between 1.5 and two gestures in more than a single context, whereas the bonobos and gorillas used more like three to four.

Overall, then, in terms of simple repertoire size and flexibility of use, there is variation among the five ape species, but not in any way that maps simply onto the ecology, social structure, or cognition of the different species.

Learning

Following Tomasello et al. (1994), we may compute concordances among the individual repertoires of different individuals of a species. For issues of social learning, the important comparison is the degree of commonality of the individuals within a single social group versus the degree of commonality of individuals across social groups, who have never been in contact with one another. Using the Kappa statistic, we looked at both within-group and between-group variability across several social groups in each species. Interestingly and importantly, the within-group and between-group variability did not differ significantly in any species — suggesting that social learning, in the form of some kind of group-specific cultural transmission, is not the major learning process at work.

Further support for this view is supplied by the fact that four of the five species had multiple individuals who used idiosyncratic gestures, presumably not learned from any other individual (the siamangs had no idiosyncratic gestures). Nevertheless, in contrast to the general pattern, there were several gestures used by multiple individuals within a particular group that were not used by the individuals in any other group (again the siamangs had none). These suggest the possibility of some form of social learning or imitation in the genesis of the gesture. For example, we found that three of four bonobos in a small captive group initiated play by *somersaulting* into one another, whereas no bonobo individuals in the other two groups we observed ever did this (Pika et al., 2005). These group-specific gestures are therefore similar to so-called 'conventional' gestures in humans, whose form and meaning are established by the convention of specific communities (e.g., *thumb-up* gesture).

It is noteworthy in terms of species differences that the major quantitative difference observed was that the overall concordance rate was lowest among chimpanzees and bonobos, reflecting more individual differences (and so perhaps more learning), and highest among gorillas, reflecting more homogeneity among individuals of the species both within and between groups. This might perhaps be related to the "fission-fusion" social structure of the two *Pan* species, in which individuals separate and reunite with one another regularly, often on a daily basis.

Adjustments for audience

Across species tactile and visual gestures were most common, each comprising from one-third to one-half of the repertoires of each species. The major differences in this regard, was that gorillas used more auditory gestures (close to one-fifth of their repertoire), including the famous *chest-beat*; chimpanzees used a fair number of auditory gestures (close to one-tenth of their repertoire), including such things as *ground-slap*; whereas orangutans and siamangs used no auditory gestures.

All five species used their visually based gestures much more often when the recipient was oriented toward them bodily (80% to 90%) than when its back was turned (10% to 20%). On the other hand, tactile gestures were used somewhat more often (about 60%) when the recipient's back was turned. It is clear that all five species understand something about how the recipient must be situated in order to receive their gesture effectively — perhaps based on understanding of what others can and cannot see (see Call & Tomasello, 2007). This may suggest that the basic social cognitive skills underlying the gesture use of the five different species are in fact quite similar.

Conclusions

The gestural modality provides a rich source of information about the nature of human and primate communication. Many researchers agree than in the vocal modality, humans use linguistic symbols whereas other primate species do not — certainly not in their natural environments. Although there is no widely agreed upon definition of linguistic symbols, at the very least they are intersubjectivity shared communicative devices used to direct attention triadically and referentially, sometimes for declarative purposes. This mode of communication clearly depends on a deep understanding of the intentional states of others, and a deep motivation to share intentional states with others as well — which seems to be especially characteristic of the human species (Tomasello et al., 2005).

Interestingly, although there are no quantitative comparisons, qualitative comparisons reveal a very similar contrast across humans and other primates in the gestural modality. Many human deictic and symbolic gestures are also used to intersubjectively to direct the attention of others referentially and for declarative purposes. Primates do not seem to use gestures in this same way. (Even apes learning language-like signs use them almost exclusively for imperative, not declarative, purposes.) However, because many of their gestures — in contrast to their vocalizations — are clearly learned and used quite flexibly, with adjustments for the attentional state of the recipient, it would seem plausible that the gestural modality of our nearest primate relatives was the modality within which symbolic communication first evolved (see also Pika, in press). The research we have reported here demonstrates interesting variability among closely related ape species in a variety of dimensions, but none of the species seems to be using either gestural or vocal symbols of the human kind — and no species stands out as doing something wildly different from the others, nor does ecology, social structure or cognition seem to make huge differences. Future research will hopefully discover potential evolutionary mechanisms by which the vocal and gestural signals of apes transformed into the linguistic and gestural symbols of human beings.

Notes

1. Apes raised in contact with humans sometimes learn to point for humans (e.g., Leavens & Hopkins, 1998), but the nature of what they are doing still seems qualitatively different from what human infants do — for example, they only point when they want something (imperatives) not when they just want to share attention (declaratives; see Tomasello & Camaioni, 1997, for a direct comparison).

2. What chimpanzees and other apes seem to do instead is to actively move around in front of the recipient before giving a visual signal (Liebal et al., 2004).

3. Examples for triadic gestures in other ape species are for instance *offer food, show object* and *move object* (Pika et al., 2003, 2005; Liebal et al., 2006).

References

Bard, Kim A. (1992). Intentional behavior and intentional communication in young free-ranging orangutans. *Child Development, 63,* 1186–1197.

Bates, Elizabeth. (1976) *Language and context: The acquisition of pragmatics.* Academic Press.

Bruner, Jerome (1981). The pragmatics of acquisition. In Werner Deutsch (Ed.), *The child's construction of language* (pp. 39–56). New York: Academic Press.

Call, Josep & Michael Tomasello (in press b). Primate gestural communication. In Marc Naguib (Ed.), *Encyclopedia of communication and language.* Amsterdam: Elsevier.

Call, Josep & Michael Tomasello (1994). Production and comprehension of referential pointing by orangutans (*Pongo pygmaeus*). *Journal of Comparative Psychology, 108*, 307–317.

Carpenter, Malinda, Katherine Nagell, & Michael Tomasello (1998). Social cognition, joint attention, and communicative competence from 9 to 15 months of age. *Monographs of the Society for Research in Child Development, Volume 255.*

Cheney, Dorothy L. & Robert Seyfarth (1990). *How monkeys see the world.* Chicago: University of Chicago Press.

Cheney, Dorothy L. & Richard W. Wrangham (1987). Predation. In Barbara B. Smuts, Dorothy L. Cheney, Robert M. Seyfarth, Richard W. Wrangham, & Thomas T. Struhsaker (Eds.), *Primate Societies* (pp. 440–451). Chicago: University of Chicago Press.

Crockford, Catherine & Christophe Boesch (2003). Context-specific calls in wild chimpanzees, *Pan troglodytes verus*: Analysis of barks. *Animal Behaviour, 66,* 115–125.

Evans, Cristopher S., Linda Evans, & Peter Marler (1993). On the meaning of alarm calls: Functional reference in an avian vocal system. *Animal Behaviour, 46,* 23–38.

Gomez, Juan C. (1990). The emergence of intentional communication as a problem-solving strategy in the gorilla. In Sue T. Parker & Kathleen R. Gibson (Eds.), *"Language" and intelligence in monkeys and apes. Comparative developmental perspectives* (pp. 333–355). New York: Cambridge University Press.

Goodall, Jane (1986). *The chimpanzees of Gombe. Patterns of behavior.* Cambridge, MA: Harvard University Press.

Kummer, Hans (1968). *Social organization of Hamadryas Baboons. A field study.* Basel: Karger.

Leavens, David A. & William D. Hopkins (1998). Intentional communication by chimpanzees: A cross-sectional study of the use of referential gestures. *Developmental Psychology, 34* (5), 813–822.

Liebal, Katja, Simone Pika, & Michael Tomasello (2006). Gestural communication of orangutans (*Pongo pygmaeus). Gesture, 6*, 1–38.

Liebal, Katja, Josep Call, & Michael Tomasello (2004). The use of gesture sequences by chimpanzees. *American Journal of Primatology, 64*, 377–396.

Lock, Andrew (1978). *The emergence of language.* In Andrew Lock (Ed.), *Action, gesture, and symbol: The emergence of language.* New York: Academic Press.

Maestripieri, Dario (1997). The evolution of communication. *Language & Communication, 17,* 269–277.

Marler, Peter (1965). Communication in monkeys and apes. In I. Devore (Ed.), *Primate Behavior: Field Studies of Monkeys and Apes* (pp. 544–584).

McGrew, William C. & Carolyne Tutin (1978). Evidence for a social custom in wild chimpanzees? *Man, N.S. 13,* 234–251.

Menzel, Emil W. (1973). Chimpanzee spatial memory organization. *Science, 182 (4115),* 943-945.

Nishida, Toshida (1980). The leaf-clipping display: A newly discovered expressive gesture in wild chimpanzees. *Journal of Human Evolution, 9,* 117–128.

Owings, Donald H. & Eugene S. Morton (1998). *Animal vocal communication: A new approach.* Cambridge: Cambridge University Press.

Owren, Michael J., Jacquelyn A. Dieter, Robert M. Seyfarth, & Dorothy L. Cheney (1992). 'Food' calls produced by adult female rhesus (*Macaca mulatta*) and Japanese (*M. fuscata*) macaques, their normally-raised offspring, and offspring cross-fostered between species. *Behaviour, 120,* 218–231.

Piaget, Jean (1952). *The origins of intelligence in children.* New York: Basic Books.

Pika, Simone, Katja Liebal, & Michael Tomasello (2003). Gestural communication in young gorillas (*Gorilla gorilla*): Gestural repertoire and use. *American Journal of Primatology, 60(3)*: 95–111.

Pika, Simone, Katja Liebal, & Michael Tomasello (2005). The gestural repertoire of bonobos (*Pan paniscus*): Flexibility and use. *American Journal of Primatology, 65,* 39–61.

Pika, Simone & John C. Mitani (2006). Referential gesturing in wild chimpanzees (*Pan troglodytes*). *Current Biology,* 16(6), 191–192.

Pika, Simone (in press). Gestures of apes and pre-linguistic human children: More similar or more different? *First Language.*

Plooij, Frans (1987). Infant-ape behavioral development, the control of perception, types of learning and symbolism. In A. Tryphon & J. Montangero (Eds.), *Symbolism and knowledge* (pp.29–58). Geneva: Jean Piaget Archives Foundation.

Slocombe, Katie E. & Klaus Zuberbühler (2005). Agonistic screams in wild chimpanzees (*Pan troglodytes schweinfurthii*) vary as a function of social role. *Journal of Comparative Psychology, 119,* 67–77.

Snowdon, Charles T., Margaret Elowson, & Rebecca S. Roush (1997). Social influences on vocal development in New World primates. In Charles T. Snowdon & Martine Hausberger (Eds.), *Social influences on vocal development* (pp. 234–248). New York, NY: Cambridge University Press.

Tanner, Joanne E. & Richard W. Byrne (1993). Concealing facial evidence of mood: Perspective-taking in a captive gorilla? *Primates, 34,* 451–457.

Tanner, Joanne E. & Richard W. Byrne (1996). Representation of action through iconic gesture in a captive lowland gorilla. *Current Anthropology, 37,* 162–173.

Tinbergen, Nico (1951). *The study of instinct.* New York: Oxford University Press.

Tomasello, Michael (1996). Do apes ape? In Bennett G. Galef & Cecilia Heyes (Eds.), *Social learning in animals: The roots of culture.* New York: Academic Press.

Tomasello, Michael & Josep Call (1994). Social cognition of monkeys and apes. *Yearbook of Physical Anthropology, 37,* 273–305.

Tomasello, Michael & Josep Call (1997). *Primate cognition.* New York: Oxford University Press.

Tomasello, Michael, Josep Call, Katherine Nagell, Raquel Olguin, & Malinda Carpenter (1994). The learning and the use of gestural signals by young chimpanzees: A trans-generational study. *Primates, 35,* 137–154.

Tomasello, Michael, Josep Call, Jennifer Warren, Thomas Frost, Malinda Carpenter, & Katherine Nagell (1997). The ontogeny of chimpanzee gestural signals: A comparison across groups and generations. *Evolution of Communication, 1,* 223–253.

Tomasello, Michael & Luigia Camaioni (1997). A comparison of the gestural communication of apes and human infants. *Human Development, 40,* 7–24.

Tomasello, Michael, Malinda Carpenter, Josep Call, Tanya Behne, & Henrike Moll (2005). Understanding and sharing intentions: The origins of cultural cognition. *Behavioral and Brain Sciences, 28,* 675–691.

Tomasello, Michael, Barbara L. George, Ann C. Kruger, Michael J. Farrar, & Andrea Evans (1985). The development of gestural communication in young chimpanzees. *Journal of Human Evolution, 14,* 175–186.

Tomasello, Michael, Deborah Gust, & Thomas Frost (1989). A longitudinal investigation of gestural communication in young chimpanzees. *Primates, 30,* 35–50.

Tomasello, Michael & Klaus Zuberbühler (2002). Primate vocal and gestural communication. In Marc Bekoff, Colin Allen, & Gordon M. Burghardt (Eds.), *The cognitive animal: Empirical and theoretical perspecitives on animal cognition.* Cambridge: MIT Press.

Tomasello, Michael & Josep Call (Eds.) (2007). *The gestural communication of apes and monkeys.* Mahwah, New York: Lawrence Erlbaum.

Whiten, Andrew & Richard W. Byrne (1988). Taking (Machiavellian) intelligence apart: Editorial. In Richard W. Byrne & Andrew Whiten (Eds.), *Machiavellian intelligence. Social expertise and the evolution of intellect in monkeys, apes, and humans* (pp. 50–65). New York: Oxford University Press.

Gestural communication in three species of macaques (*Macaca mulatta, M. nemestrina, M. arctoides*)

Use of signals in relation to dominance and social context*

Dario Maestripieri
University of Chicago

The present study compared the frequency and contextual usage of the most prominent gestural signals of dominance, submission, affiliation, and bonding in rhesus, pigtail, and stumptail macaques living in captivity. Most similarities among species were found in signals of dominance and submission and most differences in affiliative gestures and bonding patterns. Rhesus macaques have a relatively poor gestural repertoire, pigtail macaques possess conspicuous signals of affiliation and bonding, and stumptail macaques have the richest repertoire of assertive and submissive signals. The similarities and differences in the gestural repertoires of rhesus, pigtail, and stumptail macaques can be related to the intra-group social dynamics of these species as well as to their evolutionary history.

Comparisons of communication patterns across different animal species can provide evidence of the adaptive significance of signals and their phylogenetic history (e.g., Darwin, 1872; Wenzel, 1992). Since communication patterns are mainly adaptations to the social environment, in order to understand the adaptive significance and evolutionary history of the social signals observed in different species, information is needed on the social organization and behavior of these species as well as on their phylogenetic relationships (e.g. Preuschoft & van Hooff, 1996).

The genus *Macaca* includes 19 different species, which are currently subdivided into 4 distinct phyletic groups on the basis of morphological and genetic characteristics (Brandon-Jones et al., 2004; Delson, 1980; Fa, 1989; Fooden, 1980). Previous qualitative descriptions of the repertoires of facial expressions and gestures of different macaque species reported that interspecific variation is generally less pronounced in the agonistic displays (e.g., threats) than in the displays of affiliation and bonding (Bernstein, 1970; Redican, 1975; Thierry et al., 1989; van Hooff, 1967). More quantitative data and direct comparisons between different species are needed, however, before any conclusions can be made about the evolution of gestural communication in macaques.

Rhesus (*Macaca mulatta*), pigtail (*Macaca nemestrina*) and stumptail macaques (*Macaca arctoides*) belong to three different phyletic groups within the genus *Macaca* (Delson, 1980; Fooden, 1980). Pigtail macaques and related species of the *Macaca silenus* group are believed to have undergone early differentiation and dispersal, while rhesus macaques and related species of the *Macaca fascicularis* group may have differentiated and dispersed more recently (Fa, 1989). Stumptail macaques are probably related to species in the *Macaca sinica* group but seem to have undergone the most recent differentiation (Fooden, 1980).

Rhesus, pigtail, and stumptail macaques have been the focus of a number of studies involving direct interspecific comparisons of aggressive, affiliative, and maternal behavior (e.g., Bernstein et al., 1983; Butovskaya, 1993a, b; de Waal & Ren, 1988; Maestripieri, 1994; Ruehlmann et al., 1988; Weigel, 1980) and these and other studies have highlighted both similarities and differences in their social organization. Rhesus macaques live in a relatively despotic and nepotistic society characterized by high rates of aggression and spatial avoidance, and in which grooming and agonistic support mainly occur within clusters of matrilineal kin (Bernstein & Ehardt, 1985; Kaplan, 1977). The social dynamics of pigtail macaques are quite similar to those of rhesus macaques, but the lower levels of spatial avoidance, the higher reconciliation frequency, and the higher rates of approaches and grooming between pigtail females relative to rhesus (Bernstein et al., 1983; Maestripieri, 1994) suggest that the pigtail macaque society is more cohesive and conciliatory than the rhesus society. Aggression rates have been reported as similar in pigtails and rhesus (Maestripieri, 1994) or lower in the pigtails (Bernstein et al., 1983). Aggression, however, more frequently involves the participation of third individuals in pigtails than in rhesus (Bernstein et al., 1983) and post-conflict reconciliation is also frequently extended to the opponent's kin and allies (Judge, 1991). The frequency of aggression in stumptails has been reported as higher than in rhesus and pigtails (Butovskaya 1993a, b; de Waal & Ren 1988; Weigel, 1980). Although some authors reported that stumptail aggression only rarely escalates to serious biting (de Waal & Ren, 1988), according to others biting is as frequent as in rhesus and more frequent than in pigtails (Bernstein, 1980; Ruehlmann et al., 1988). Stumptail macaques also exhibit higher rates of proximity, contact, huddling, and grooming than rhesus and pigtails (Bernstein, 1980; Butovskaya, 1993a; de Waal & Ren, 1988; Maestripieri, 1994). The co-existence of high intragroup aggression and high cohesion in stumptail macaques could be related to the retention of supernumerary adult males in the social group for competition with other groups or protection from predators (e.g. Bertrand, 1969; Estrada et al., 1977). Stumptail males have been reported as being twice as aggressive as rhesus males and four times as aggressive as pigtail males (Ruehlmann et al., 1988). Stumptail males are also significantly larger and more aggressive than females and easily overpower them also in sexual interactions, where forced copulations are not unusual (Bernstein et al., 1983; Bertrand, 1969; Ruehlmann et al., 1988). Moreover, post-copulatory tieing with females, prolonged mate guarding, and surreptitious copulations suggest intense mating and sperm competition between stumptail males (Brereton, 1993; Estep et al., 1988).

Variation in social organization between rhesus, pigtail, and stumptail macaques should be accompanied by differences in social communication. Previous studies investigating the use of nonvocal signals in each of these three species and comparing the size of their gestural repertoire suggested that this is indeed the case (Maestripieri, 1996a, b, 1999; Maestripieri & Wallen, 1997). The present study expands the previous comparative investigation of gestural communication in rhesus, pigtail, and stumptail macaques by investigating the frequency of occurrence of nonvocal signals and their use in relation to dominance rank and social context. The findings are discussed in light of information on social organization and phylogenetic relationships between rhesus, pigtail and stumptail macaques to elucidate the adaptive significance and evolution of gestural communication in these species.

Method

All study subjects lived in social groups housed in large outdoor compounds at the Field Station of the Yerkes National Primate Research Center in Lawrenceville, Georgia (U.S.A.). Group size and composition were similar to those in the wild. The rhesus group consisted of 2 adult males and 26 adult females with their subadult, juvenile, and infant offspring. The pigtail group consisted of 5 adult males and 28 adult females with their offspring, and the stumptail group consisted of 8 adult males and 17 adult females with their offspring. The dominance hierarchy within each group was determined on the basis of data on aggression and spatial displacements recorded during previous studies.

Each group was observed for 100 hr during an 8-month period, between August 1994 and April 1995. Data were collected during 30-min observation sessions randomly distributed between 0800 and 1900 hr. Observations were made from a tower that provided an unrestricted view of the entire compound. All data were collected by the same observer using a tape-recorder and then transferred into a computer. Data were collected with the behavior sampling method, i.e., the observer watched the whole group and recorded each occurrence of a particular type of behavior, together with other related behaviors and details of the individuals involved.

Fifteen facial expressions, hand gestures, and body postures (collectively referred to as gestures) were selected for observation on the basis of previous studies and preliminary observations of the study subjects. The operational definitions of these signals are presented in Table 1. Since threat and play displays such as the "staring open-mouth face" and the "relaxed open-mouth face" (van Hooff, 1967) are remarkably similar in structure and contextual usage in these species, they were not included in this comparative study. Behavioral sequences involving the signals were recorded only when the behavior preceding the signal (e.g. approach or aggression) was actually observed, and were followed until the end (e.g., when

Table 1. Behavioral definitions of gestures

Gesture	Definition
Lip-Smack (LS)	Rapid opening and closing of the mouth and lips, such that when the lips close they make an audible smacking sound.
Pucker (PC)	The lips are compressed and protruded, the eyebrows, forehead and ears are retracted.
Teeth-Chatter (TC)	The mouth is rapidly opened and closed and the lips are retracted, exposing the teeth.
Bared-Teeth (BT)	The mouth is closed and the lips and lip corners are retracted so that the teeth are exposed in a white band.
Eye-Brows (EB)	The scalp and brow are retracted and the mouth is open.
Touch-Face (TF)	One hand is extended to touch the face of another individual while standing or sitting in front of it.
Touch-Genitals (TG)	Manipulation of the genitals of another individual without olfactory inspection.
Present (PR)	The tail is raised to expose the genitals.
Hip-Touch (HT)	Brief touch of the hindquarters of another individual with one or both hands without putting arms around.
Hip-Clasp (HC)	The hindquarters of another individual are clasped with both arms, usually in the sitting position.
Mount (MT)	Mount with or without foot-clasp but with no intromission or thrusts.
Present-Arm (PA)	One arm or hand is extended across the face of another individual to be bitten.
Mock-Bite (MB)	Gripping another individual's skin with the teeth, slowly, without roughness, for several seconds.
Face-Inspection (FI)	Close inspection of the face of another individual, usually staring into its eyes for several seconds, while the other individual freezes (not recorded during feeding).
Embrace (EM)	Ventral embrace with both arms around the torso of another individual, in the sitting position and kneading the partner's fur or flesh.

two individuals were more than 5 m apart from one another and did not further interact for 10–20 s). The occurrence of any interaction between the sender and receiver of the signal as well as the behavior of any other individuals participating in the interaction were recorded. Other behavioral interactions recorded during the observation sessions included approaches and leaves within arm's reach, contact, grooming, aggression (threats, bites, chases), avoidance, vocalizations (screams and grunts), play, and infant handling.

The occurrence of signals was compared among the three species in relation to dominance rank and various social contexts including after receiving aggression, in response to an approach or another signal, unsolicited (i.e. in conjunction with a spontaneous approach), and before an affiliative interaction such as contact, grooming or play. These contexts were selected for analysis because previous studies showed that they are often associated with communicative interactions in all three species (Maestripieri, 1996a, b; Maestripieri & Wallen, 1997).

Interspecific comparisons in the frequency of gestures were conducted with a one-way analysis of variance (ANOVA). Comparisons of the contexts of occurrence of gestures were conducted with two-way ANOVAs for repeated measures. Bonferroni-Dunn tests were used as post-hocs. All statistical tests are two-tailed. Although statistical analyses of contextual usage of gestures used data points for all individuals, data are presented in terms of percentage scores.

Results

Figure 1 shows the frequency of occurrence of all gestures in the three species. A previous analysis showed that the frequency of gestures (all gestures combined) was significantly different in the three species, being lowest in rhesus macaques, highest in stumptails, and intermediate in pigtails (Maestripieri, 1999). In rhesus macaques, only 4 gestures were displayed with a frequency equal to or greater than 1 event per individual, compared to 8 gestures in pigtail macaques and 12 gestures in stumptail macaques.

Frequency of individual gestures

Lipsmack (LS), Bared-Teeth (BT), Present (PR), and Mount (MT) were frequent (≥ 1 event per individual) in all three species but their frequency of occurrence was significantly different (LS: $F_{2,178} = 28.05$, $p < 0.0001$; BT: $F_{2,178} = 10.51$, $p = 0.0001$; PR: $F_{2,178} = 57.15$, $p < 0.0001$; MT: $F_{2,178} = 3.11$, $p < 0.05$). Lip-Smack was more frequent in pigtails than in rhesus ($p < 0.0001$) and stumptails ($p < 0.0001$),

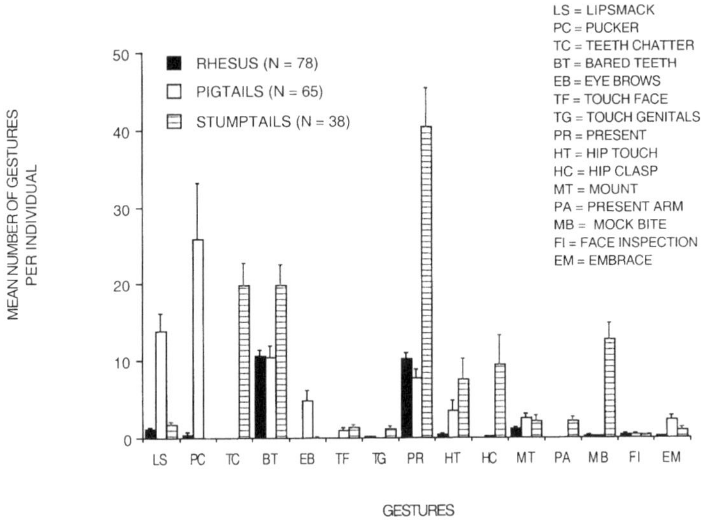

Figure 1. Mean (+ SEM) number of gestures per individual observed in the three species (Modified after Maestripieri 1999).

whereas there was no significant difference between rhesus and stumptails. Bared-Teeth and Present were more frequent in stumptails than in rhesus ($p < 0.0001$) and pigtails ($p < 0.0001$), with no significant difference between rhesus and pigtails. Finally, Mount was more frequent in rhesus than in pigtails ($p = 0.02$), with no significant differences between rhesus and stumptails, or between pigtails and stumptails.

Hip-Touch (HT), Mock-Bite (MB), Embrace (EM), Touch-Face (TF), and Touch-Genitals (TG) were observed in all three species, but only infrequently (< 1 event per individual) in one or two species. The frequency of occurrence of these gestures was significantly different in the three species (HT: $F_{2,178} = 6.17$, $p < 0.01$; MB: $F_{2,178} = 68.51$, $p < 0.0001$; EM: $F_{2,178} = 92.88$, $p < 0.0001$; TF: $F_{2,178} = 8.04$, $p < 0.001$; TG: $F_{2,178} = 12.28$, $p < 0.0001$). Hip-Touch and Touch-Face were more frequent in pigtails and stumptails than in rhesus (all values $p < 0.01$; no significant differences between pigtails and stumptails), Mock-Bite and Touch-Genitals were more frequent in stumptails than in rhesus ($p < 0.0001$) and pigtails ($p < 0.001$; no significant differences between rhesus and pigtails), and Embrace was more frequent in pigtails than in rhesus ($p < 0.0001$) and stumptails ($p < 0.01$; no significant difference between rhesus and stumptails).

Pucker (PC) was common in pigtail macaques, but very rare in rhesus and nonexistent in stumptails. In contrast, Teeth-Chatter (TC), Present Arm (PA), and Hip-Clasp (HC) were common in the stumptails but virtually absent in the other two species. Finally, Face-Inspect (FI) was very infrequent (< 1 event per indi-

Table 2. Interspecific comparisons in the frequency of occurrence of gestures, the extent to which they are mostly directed up or down the hierarchy and their contextual use

Gesture	Frequency	Hierarchy		Aggression	Approach	Unsolicited	Pre-affiliation	Post-Present	Pre-Mount
Lip-Smack	P > R = S	(up)	P > R = S	P > S > R	P > S > R	R = S > P	R = S > P	—	—
Bared-Teeth	S > R = P	(up)	S > R = P	R = S > P	P > R = S	S > R = P	P > R = S	—	—
Present	S > R = P	(up)	S > R = P	R > P = S	R > P = S	S > P > R	P > S > R	—	—
Mount	R > P = S	(down)	R = P = S	—	—	R = P = S	R = P > S	R = S > P	—
Hip-Touch	P = S > R	(down)	R = P = S	—	—	R = P > S	R = P = S	S > R = P	R = P > S
Mock-Bite	S > R = P	(down)	R = P = S	—	—	—	—	—	—
Embrace	P > R = S	—	R = P = S	—	—	—	—	—	—
Touch-Face	P = S > R	(up)	P = S > R	—	—	—	—	—	—
Touch-Gen.	S > R = P	—	R = P = S	—	—	—	—	—	—
Pucker	P > R = S	—							
Teeth-Chatter	S > R = P	(up)							
Present-Arm	S > R = P	(up)							
Hip-Clasp	S > R = P	(down)							
Face-Inspect	—	(down)							

R = rhesus; P = pigtails; S = stumptails

vidual) in all three species. Table 2 summarizes the results of interspecific comparisons in the frequency of occurrence of all gestures. The gestures that were virtually unique to one species, or infrequent in all species, were not statistically compared among the species. The occurrence of these gestures will be discussed on basis of data analyses reported elsewhere (Maestripieri, 1996a, b; Maestripieri & Wallen, 1997).

Effects of dominance hierarchy

Lip-Smack, Bared-Teeth, and Present were displayed by subordinates to dominants more than vice versa in all three species (LS: $F\ 1{,}180 = 40.70$, $p < 0.0001$; BT: $F\ 1{,}180 = 161.37$, $p < 0.0001$; PR: $F\ 1{,}180 = 112.11$, $p < 0.0001$).

The three species, however, differed significantly in the extent to which these gestures were directed up the hierarchy (LS: $F\ 2{,}178 = 29.14$, $p < 0.0001$; BT: $F\ 2{,}178 = 8.50$, $p < 0.001$; PR: $F\ 2{,}178 = 47.90$, $p < 0.0001$). Specifically, the proportion of Lip-Smack directed up the hierarchy was significantly higher in pigtails than in rhesus ($p < 0.0001$) and stumptails ($p < 0.0001$), with no significant difference between rhesus and stumptails. The proportion of Bared-Teeth and Present directed up the hierarchy was significantly higher in stumptails than in rhesus (BT: $p < 0.01$; PR: $p < 0.001$), and in pigtails (BT: $p < 0.01$; PR: $p < 0.001$), with no significant difference

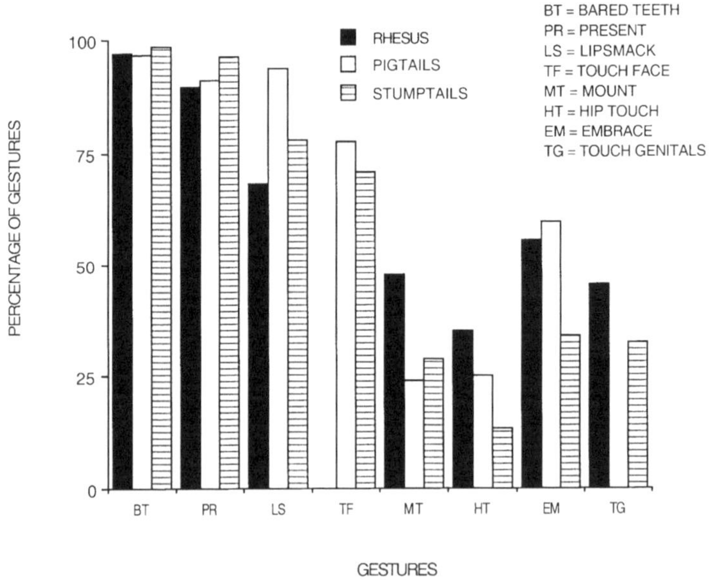

Figure 2. Percentage of gestures directed up the hierarchy in the three species. Only gestures occurring in at least two of the three species are shown.

between rhesus and pigtails. Touch-Face was mostly displayed by subordinates in both pigtails and stumptails ($F_{1,35} = 5.97$, $p < 0.05$), with no significant difference between these species. The two events observed among rhesus macaques were displayed by mothers to their newborn infants. Mount, Hip-Touch, and Mock-Bite were displayed by dominants more than by subordinates in all three species (MT; $F_{1,134} = 9.67$, $p < 0.01$; HT: $F_{1,108} = 6.96$, $p < 0.01$; MB: $F_{1,50} = 16.09$, $p < 0.001$) and there were no significant differences in the extent to which these behaviors were directed down the hierarchy. Embrace and Touch-Genitals occurred irrespective of dominance rank in all species. Figure 2 illustrates the percentage of gestures directed up the hierarchy (i.e. from subordinates to dominants) in the three species.

Contexts of occurrence

The occurrence of Lip-Smack, Bared-Teeth, and Present was compared in four social contexts: after receiving aggression, in response to an approach (in most cases, by a dominant individual), in conjunction with a spontaneous approach, and prior to affiliation. The first three contexts are mutually exclusive but the fourth can overlap with any of them (e.g., individuals can display a signal in response to aggression and then engage in affiliative behavior). There were significant interspecific differences in the occurrence of the three signals in the four contexts (aggression, LS: $F_{2,121} = 12.43$, $p < 0.0001$, BT: $F_{2,169} = 46.46$, $p < 0.0001$; PR: $F_{2,169} = 71.32$, $p < 0.0001$; approach, LS: $F_{2,121} = 10.17$, $p < 0.0001$; BT: $F_{2,169} = 58.03$, $p < 0.0001$; PR: $F_{2,169} = 49.05$, $p < 0.0001$; unsolicited, LS: $F_{2,121} = 20.86$, $p < 0.0001$; BT: $F_{2,169} = 11.50$, $p < 0.0001$; PR: $F_{2,169} = 71.24$, $p < 0.0001$; pre-affiliation, LS: $F_{2,121} = 14.68$, $p < 0.0001$; BT: $F_{2,169} = 13.09$, $p < 0.0001$; PR: $F_{2,169} = 50.29$, $p < 0.0001$). Pigtails displayed Lip-Smack more frequently after receiving aggression and in response to an approach than rhesus (aggression, pigtails: 14.58%, rhesus: 1.96%, $p < 0.0001$; approach, pigtails: 26.39%, rhesus: 1.96%, $p < 0.0001$) and stumptails (aggression: 8.82%, $p < 0.01$; approach: 20.58%, $p < 0.05$). Stumptails displayed Lip-Smack in response to aggression and approach more than rhesus ($p < 0.01$). In contrast, rhesus and stumptails displayed Lip-Smack with a spontaneous approach more than pigtails (rhesus: 52.94%; pigtails: 19.59%; stumptails: 41.17%; rhesus-pigtails: $p < 0.001$, stumptails-pigtails: $p < 0.001$, rhesus-stumptails, NS). Lip-Smack was more likely to be followed by affiliation in rhesus (60.78%) and stumptails (57.35%) than in pigtails (18.04%; rhesus-pigtails: $p < 0.001$, stumptails-pigtails: $p < 0.001$, rhesus-stumptails, NS). In the pigtails, Bared-Teeth was less likely to occur after receiving aggression (41.08%), more likely to occur in response to an approach (49.23%), and more likely to be followed by affiliation (6.11%) than in rhesus (aggression: 64.76%; approach: 26.73%; affiliation: 2.18%;

all p values < 0.001) and stumptails (aggression: 67.99%; approach: 20.58%; affiliation: 2.39%; all p values < 0.001). Rhesus and stumptails did not differ significantly in any of these contexts. Bared-Teeth, however, was displayed unsolicited by stumptails (3.45%) more than by rhesus (2.18%; $p < 0.05$) and pigtails (2.88%; $p < 0.05$; rhesus-pigtails, NS). Rhesus displayed Present in response to aggression (35.64%) and approach (39.16%) more than pigtails (aggression: 9.74%; approach: 20.89%; $p < 0.001$) and stumptails (aggression: 6.58%; approach: 22.96; %; $p < 0.01$; pigtails-stumptails, NS). Stumptails displayed unsolicited Present (69.73%) more than rhesus (22.32%; $p < 0.0001$) and pigtails (48.68%; $p < 0.01$). Pigtails displayed unsolicited Present more than rhesus ($p < 0.05$). In pigtails, Present was more likely to be followed by affiliation (23.93%) than in rhesus (3.13%; $p < 0.0001$) and stumptails (10.57%; $p < 0.01$). Present was more likely to be followed by affiliation in stumptails than in rhesus ($p < 0.01$).

Mount was compared in the following contexts: unsolicited (i.e, one individual approached another one and mounted him/her without any prior interactions between them), in response to Present, and before affiliation. There were no significant differences among species in the occurrence of unsolicited Mount (rhesus: 17.86%; pigtails: 18.18%; stumptails: 20.51%), but species differed significantly in the proportion of Mount that occurred in response to Present (F 2,83 = 8.46, $p < 0.001$) and prior to affiliation (F 2,83 = 6.02, $p < 0.01$). Specifically, rhesus and stumptails were more likely to display Mount in response to Present (rhesus: 58.33%; stumptails: 51.28%) than pigtails (33.33%; $p < 0.01$; rhesus-stumptails, NS), and in rhesus and pigtails Mount was more likely to be followed by affiliation (rhesus: 30.95%; pigtails: 33.33%) than in stumptails (5.13%; $p < 0.01$; rhesus-pigtails, NS).

Hip-Touch differed among species in the extent to which it was displayed unsolicited (F 2,74 = 7.69, $p < 0.001$) or in response to Present (F 2,74 = 4.86, $p = 0.01$). Hip-Touch also differed in the extent to which it was followed by Mount (F 2,74 = 3.98, $p < 0.05$) but not by affiliation. Hip-Touch was more frequently unsolicited in rhesus (64.86%) and pigtails (88.39%) than in stumptails (19.50%; $p < 0.001$), and occurred more frequently in response to Present in stumptails (64.18%) than in rhesus (27.02%; $p < 0.05$) and pigtails (8.48%; $p < 0.01$). In rhesus and pigtails, Hip-Touch was also followed by Mount (rhesus: 16.21%; pigtails: 19.19%) more frequently than in stumptails (4.96%; $p < 0.05$).

The frequency of Mock-Bite, Embrace, Touch-Face and Touch-Genitals was too low in some species for a quantitative contextual analysis. Mock-Bite was often displayed after attacking another individual (rhesus: 40%; pigtails: 35.71%; stumptails: 57.20%) and often followed by Bared-Teeth. Embrace was mostly displayed by females (rhesus: 66.67%; pigtails: 97.43%; stumptails: 84.21%) and was often followed by huddling or grooming (rhesus: 77.78%; pigtails: 71.79%: stumptails: 42.10%). Touch-Face was often displayed in conjunction with facial expressions such as Bared-Teeth, Lip-Smack, Pucker, or Teeth-Chatter (rhesus: 100%; pigtails: 79.68%: stumptails: 82.69%). Touch-Genitals was mostly exchanged between males (rhesus: 100%; pigtails: 100%: stumptails: 74.42%).

Species-specific or infrequent gestures

Pucker was the most frequent gesture observed in pigtail macaques. Pucker was never observed among stumptails and only on a few occasions among rhesus. In pigtails, Pucker was displayed by both males and females independent of their dominance rank and in a variety of social contexts, including mating, grooming, and interactions with infants. Eye-Brows was also unique to pigtail macaques, where it was frequently exchanged between males, irrespective of their dominance rank, in conjunction with approach-retreat interactions, Hip-Touch, grunts, and occasionally brief bouts of play. Eye-Brows occurred in conjunction with agonistic support and was often followed by affiliation.

Teeth-Chatter, Present Arm, and Hip-Clasp were virtually unique to stumptail macaques. Teeth-Chatter was mostly directed up the hierarchy and often associated with Hip-Touch, Hip-Clasp, Mount and Embrace between females. Present-Arm was mostly displayed by subordinates and followed by Mock-Bite by dominants. Hip-Clasp was mostly displayed by the alpha male, and occurred in contexts similar to those of Hip-Touch, and primarily in response to Present. Unlike Hip-Touch, most Hip-Clasp was directed to juveniles and infants who solicited this behavior in the presence of an external threat to the group or during disputes with other juveniles. Face-Inspect occurred with a frequency lower than 1 event per in individual in all three species and was typically displayed by dominants after they approached subordinates. It elicited freezing in the subordinate or a submissive signal such as Bared-Teeth.

Discussion

Several main findings emerge from this comparative study of gestural communication in macaques. First, the gestural repertoire of rhesus macaques is generally poor in comparison to that of pigtail macaques, and especially that of stumptail macaques. Rhesus macaques exhibit fewer signals and use some of them with a lower frequency than the other species (Maestripieri, 1999). Second, most communication in these three species appears to revolve around issues of dominance and submission (Maestripieri, 1996a, b; Maestripieri & Wallen, 1997). Third, most similarities in the gestural repertoires of rhesus, pigtail and stumptail macaques were found in submissive and assertive signals and the greatest variability in communicative patterns related to affiliation and bonding. Even among the submissive and assertive signals, however, there are quantitative differences among the species, as submissive and assertive signals were more numerous and more frequent among stumptails than among rhesus and pigtails.

Bared-Teeth, Present, and Lip-Smack were among the most frequent signals occurring in the three species and, in all species, they were strictly directed up the hierarchy. In contrast, Hip-Touch and Mount (and in the stumptails also Mock-Bite) were generally directed down the hierarchy. Other gestures, which were limited to one or two species and did not have a clear relationship with dominance, were Pucker, Embrace, and Touch-Genitals.

In rhesus macaques Bared-Teeth and Present were mainly displayed in response to aggression or an approach by a dominant individual and rarely followed by affiliation. Although rhesus macaques have few affiliative signals relative to the other species, Lip-Smack appears to have a stronger affiliative component in rhesus than in the other species, as this signal was often unsolicited and followed by affiliation. In pigtail macaques, Bared-Teeth and Present occurred in contexts similar to those of rhesus and stumptails, but they were more frequently followed by affiliation. In pigtails, however, the contextual use of Lip-Smack was more similar to that of Bared-Teeth and Present than in the other species. Pucker was the most frequent gesture observed in pigtail macaques. Previous studies showed that Pucker is used to coordinate and facilitate the occurrence of mating, grooming, and interactions with infants (Maestripieri, 1996a; see also Jensen & Gordon, 1970). Pigtail macaques also exhibit frequent bonding patterns such as Embrace and Eye-Brows. In stumptail macaques, Bared-Teeth and Present were very frequent, mostly unsolicited, and strongly directed up the hierarchy, suggesting that they serve an appeasing function. Stumptail macaques possess further submissive gestures such as Present-Arm, Teeth-Chatter, and Touch-Face. Furthermore, in this species, Mount was more likely to occur in response to Present and less likely to be followed by affiliation than in the other species, suggesting that this behavior, along with Hip-Touch, has a strong assertive component. Stumptail macaques also have bonding patterns such as Embrace, Hip-Clasp, and Touch-Genitals, some of which may serve a reassurance or protection function.

It may be argued that whereas the richness of the dominance/ submission communicative repertoire reflects the potential for competition and conflict within groups, affiliative signals and bonding patterns probably reflect the need for intragroup cohesion and cooperation for defense against predators or competition with other groups. In a despotic and nepotistic society like that of rhesus macaques there may be little pressure to develop a sophisticated system of affiliative signals and bonding patterns. Maintenance of group structure and coordination of behavior between individuals can be effectively achieved if a few unequivocal indicators of differences in dominance are recognized and if unrelated or distantly-ranked individuals simply avoid each other (Maestripieri, 1999). In pigtail macaques, instead, complex dynamics of intragroup cooperation and high levels of social tolerance appear to have led to the evolution of intense affiliative communication and bonding patterns. The variety of assertive and submissive signals observed in stumptail macaques suggests a great potential for intraspecific conflict. Communication of dominance and submission, however, is also frequently accompanied by expressions of reassurance and bonding, suggesting the need for intragroup cohesion and cooperation.

Submissive signals such as Bared-Teeth and Present are remarkably similar in rhesus, pigtail, and stumptail macaques suggesting that these signals (probably along with threat displays, the play-face, Lip-Smack, and Mount) were present in the ancestor of these species. In fact, these signals also appear in most, if not all, of the other African *Cercopithecidae* (Andrew, 1963; van Hooff, 1967; Redican, 1975). Pucker is a common gesture in pigtail and liontail macaques (*Macaca silenus*; Lindburg et al., 1985; Johnson, 1985) but rare in rhesus and longtail macaques (*Macaca fascicularis*; Shirek-Ellefson, 1972) and absent in the stumptails, suggesting that it may be a relatively ancestral signal that has been conserved in the *silenus* group but partially lost in other species. Ventro-ventral Embrace has been reported in species of all four phyletic groups of macaques (Thierry, 1984), and especially in the *silenus* group (Dixson, 1977; Skinner & Lockard, 1979; Thierry, 1984) and in *Macaca fascicularis* (Shirek-Ellefson, 1972), which is closely related to rhesus macaques. It seems likely, therefore, that Embrace is a relatively ancestral pattern that has become very infrequent in rhesus macaques. Finally, Teeth-Chatter has been reported in Barbary macaques (*Macaca sylvanus*; van Hooff, 1967), which are believed to be the most ancestral macaque species, and in macaque species of the *sinica* group (e.g. bonnet, Tibetan, and assamese macaques), which are probably closely related to stumptail macaques (Fooden, 1980). This suggests that Teeth-Chatter evolved relatively early in macaques, was retained in Barbary macaques and species of the *sinica* group including stumptail macaques, and was lost in other species such as pigtail and rhesus. Different macaque species, however, may have independently evolved Teeth-Chatter from other signals such Bared-Teeth and Lip-Smack (see van Hooff, 1967).

Signals such as Eye-Brows, Teeth-Chatter, Hip-Clasp, Present-Arm, and Mock-Bite may have evolved independently in some macaque species. Eye-Brows has also been reported in *Macaca silenus* (Johnson, 1985; Skinner & Lockard, 1979), suggesting that it may have evolved independently species of the *silenus* group. Hip-Clasp and especially Present-Arm and Mock-Bite are behavior patterns virtually unique to stumptail macaques. Hip-Clasp and perhaps also Touch-Genitals between stumptail adults probably develop from ritualized interactions between adults and infants in which adults lift the infant's hindquarters and hold them briefly while manipulating the infant's genitals and teeth-chattering (this interaction has been referred to as "bridging"; Bertrand, 1969; see Ogawa, 1995, for *Macaca thibetana*).

In conclusion, this study suggests that the similarities and differences in the gestural repertoires of rhesus, pigtail, and stumptail macaques can be related to the intragroup social dynamics of these species as well as to their evolutionary history. Future studies should extend the comparison of communication patterns to other species of macaques and discuss their findings in relation to the phylogeny and social evolution of this primate genus.

Note

This work was supported in part by NIH grant RR-00165 awarded to the Yerkes National Primate Research Center. The Yerkes Center is fully accredited by the American Association for Accreditation of Laboratory Animal Care.

References

Andrew, Richard J. (1963). The origin and evolution of the calls and facial expressions of the primates. *Behaviour*, *20*, 1–109.

Bernstein, Irwin S. (1970). Some behavioral elements of the *Cercopithecoidea*. In John H. & Prue H. Napier (Eds.), *Old World monkeys. Evolution, systematics and behavior* (pp. 263–295). New York: Academic Press.

Bernstein, Irwin S. (1980). Activity patterns in a stumptail macaque group. *Folia Primatologica*, *33*, 20–45.

Bernstein, Irwin S. & Carolyn L. Ehardt (1985). Agonistic aiding: Kinship, rank, age and sex influences. *American Journal of Primatology*, *8*, 37–52.

Bernstein, Irwin S., Lawrence Williams, & Marcy Ramsay (1983). The expression of aggression in Old World monkeys. *International Journal of Primatology*, *4*, 113–125.

Bertrand, Mireille (1969). *The behavioral repertoire of the stumptail macaque*. Basel: Karger.

Brandon-Jones, Douglas, Ardith Eudey, Thomas Geissmann, Colin P. Groves, Donald J. Melnick, Juan Carlos Morales, Myron Shekelle, & Caro-Beth Stewart (2004). Asian primate classification. *International Journal of Primatology*, *25*, 97–164.

Brereton, Alyn (1993). Evolution of the sociosexual pattern of the stumptail macaque (*Macaca arctoides*). *Folia Primatologica, 61*, 43–46.

Butovskaya, Marina (1993a). Kinship and different dominance styles in groups of three species of the genus *Macaca (M. arctoides, M. mulatta, M. fascicularis). Folia Primatologica, 60*, 210–224.

Butovskaya, Marina (1993b). Intrusion into agonistic encounters in 3 species of genus *Macaca* (*Macaca arctoides, M. mulatta, M. fascicularis*) with reference to different dominant styles. *Primate Report, 37*, 41–50.

Darwin, Charles (1872). *The expression of the emotions in man and animals*. London: Murray.

Delson, Eric (1980). Fossil macaques, phyletic relationships and a scenario of development. In Donald G. Lindburg (Ed.), *The macaques. Studies in ecology, behavior, and evolution* (pp. 10–30). New York: Van Nostrand Reinhold.

de Waal, Frans B. M. & Ren Mei Ren (1988). Comparison of the reconciliation behavior of stumptail and rhesus macaques. *Ethology, 78*, 129–142.

Dixson, Alan F. (1977). Observations on the displays, menstrual cycles and sexual behaviour of the "Black Ape" of Celebes (*Macaca nigra*). *Journal of Zoology, 182*, 63–84.

Estep, Daniel Q., Kees Nieuwenhuijsen, Katherine E. Bruce, Karel J. de Neef, Paul A. Walters, Suzanne C. Baker, & Koos A. Slob (1988). Inhibition of sexual behaviour among subordinate stumptail macaques (*Macaca arctoides*). *Animal Behaviour, 36*, 854–864.

Estrada, Alejandro, Rosamond Estrada, & Frank Ervin, F. (1977). Establishment of a free-ranging colony of stumptail macaques (*Macaca arctoides*): I. Social relations. *Primates, 18*, 647–676.

Fa, John E. (1989). The genus *Macaca*: A review of taxonomy and evolution. *Mammal Reviews, 19*, 45–81.

Fooden, Jack (1980). Classification and distribution of living macaques (*Macaca* Lacépède, 1799). In Donald G. Lindburg (Ed.), *The macaques. Studies in ecology, behavior, and evolution* (pp. 1–9). New York: Van Nostrand Reinhold.

Jensen, Gordon D. & Betty N. Gordon (1970). Sequences of mother-infant behavior following a facial communicative gesture of pigtail monkeys. *Biological Psychology, 2*, 267–272.

Johnson, Pearce C. (1985). Notes on the ethogram of captive lion-tailed macaques. In Paul G. Heltne (Ed.), *The lion-tailed macaque. Status and conservation* (pp. 239–263). New York: Alan Liss.

Judge, Peter G. (1991). Dyadic and triadic reconciliation in pigtail macaques (*Macaca nemestrina*). *American Journal of Primatology, 23*, 225–237.

Kaplan, Jay R. (1977). Patterns of fight interference in free-ranging rhesus monkeys. *American Journal of Physical Anthropology, 47*, 279–287.

Lindburg, Donald G., S. Shideler, & H. Fitch (1985). Sexual behavior in relation to time of ovulation in the lion-tailed macaque. In Paul G. Heltne (Ed.), *The lion-tailed macaque. Status and conservation* (pp. 131–148). New York: Alan Liss.

Maestripieri, Dario (1994). Mother-infant relationships in three species of macaques (*Macaca mulatta, M. nemestrina, M. arctoides*). II. The social environment. *Behaviour, 131*, 97–113.

Maestripieri, Dario (1996a). Gestural communication and its cognitive implications in pigtail macaques (*Macaca nemestrina*). *Behaviour, 133*, 997–1022.

Maestripieri, Dario (1996b). Social communication among captive stumptail macaques (*Macaca arctoides*). *International Journal of Primatology, 17*, 785–802.

Maestripieri, Dario (1999). Primate social organization, gestural repertoire size size, and communication dynamics: A comparative study of macaques. In Barbara J. King (Ed.), *The origins of language: What nonhuman primates can tell us* (pp. 55–77). Santa Fe, NM: The School of American Research.

Maestripieri, Dario & Kim Wallen (1997). Affiliative and submissive communication in rhesus macaques. *Primates, 38*, 127–138.

Ogawa, Hideshi (1995). Recognition of social relationships in bridging behavior among Tibetan macaques (*Macaca thibetana*). *American Journal of Primatology, 35*, 305–310.

Preuschoft, Signe & Jan A. R. A. M. van Hooff (1996). Homologizing primate facial displays: A critical review of methods. *Folia Primatologica, 65*, 121–137.

Redican, William K. (1975). Facial expressions in nonhuman primates. In Leonard A. Rosenblum (Ed.), *Primate behavior. Developments in field and laboratory research* (vol. 4, pp. 103–194). New York: Academic Press.

Ruehlmann, Thomas E., Irwin S. Bernstein, Thomas P. Gordon, & Peter Balcaen (1988). Wounding patterns in three species of captive macaques. *American Journal of Primatology, 14*, 125–134.

Shirek-Ellefson, Judith (1972). Social communication in some Old World monkeys and gibbons. In Phyllis Dolhinow (Ed.), *Primate patterns* (pp. 297–311). New York: Holt, Rinehart & Winston.

Skinner, Samuel W. & Joan S. Lockard (1979). An ethogram of the liontailed macaque (*Macaca silenus*) in captivity. *Applied Animal Ethology, 5*, 241–256.

Thierry, Bernard (1984). Clasping behavior in *Macaca tonkeana*. *Behaviour, 89*, 1–28.

Thierry, Bernard, Christine Demaria, Signe Preuschoft, & Christine Desportes (1989). Structural convergence between silent bared-teeth display and relaxed open-mouth display in the tonkean macaque (*Macaca tonkeana*). *Folia Primatologica, 52*, 178–184.

van Hooff, Jan A. R. A. M. (1967). The facial displays of the Catarrhine monkeys and apes. In Desmond Morris (Ed.), *Primate ethology* (pp. 7–68). London: Weidenfield.

Weigel, Robert M. (1980). Dyadic spatial relationships in pigtail and stumptail macaques: A multiple regression analysis. *International Journal of Primatology, 1*, 287–321.

Wenzel, John W. (1992). Behavioral homology and phylogeny. *Annual Reviews of Ecology & Systematics*, 23, 361–381

Multimodal concomitants of manual gesture by chimpanzees (*Pan troglodytes*)

Influence of food size and distance

David A. Leavens and William D. Hopkins
University of Sussex / Yerkes National Primate Research Center
& Berry College

It is well-established that chimpanzees vocalize more in the presence of relatively large amounts of food. The present study administered four trials in random order to each of 20 chimpanzees: (1) small piece of fruit, placed near to cage (~30 cm.), (2) large piece of fruit, placed near to cage, (3) small piece of fruit, placed far from cage (~130 cm.), and (4) large piece of fruit, placed far from cage. On arrival of an experimenter, the chimpanzees not only vocalized more in the presence of the large piece of fruit, confirming previous studies' findings, but also exhibited more multimodal behavior (vocalizations, manual gestures, and gaze alternation between the food and the experimenter), which extends previous research. More gaze alternation was exhibited to food placed more peripherally. Arousal may be indexed in this species by the number of modalities in which they communicate.

Both the propensity of captive chimpanzees (*Pan troglodytes*) to vocalize and their calling rates, in the presence of food, are related to the amount and divisibility of that food (Hauser & Wrangham, 1987; Hauser, Teixidor, Field, & Flaherty, 1993; reviewed by Hauser, 1996). In short, the more food is available, or the more divisible food is, the more likely are chimpanzees to vocalize. Thus, ecological factors seem to influence chimpanzee vocal communication, although as Hauser (1996) noted, it is not clear whether this vocal behavior has a semantic function that refers to quantity or is a direct reflection of the amount of arousal elicited by food arrays of various physical dimensions (these are not necessarily mutually exclusive interpretations). For example, for a food with given volume, if that food is presented as numerous pieces, rather than as a single, large, piece, it might appear to constitute more food (e.g., by subtending a larger visual angle). Leavens and Hopkins (1998) speculated that the number of modalities in which chimpanzees communicate about unreachable food might directly index their level of arousal.

If the amount of visible food is proportional to the level of arousal, then one might predict not only more vocal production, but also an increase in the display of signals in multiple sensory domains as a function of the amount of food.

With respect to multimodal communication by apes, previous research has demonstrated that chimpanzees display communication in different sensory modalities as a function of differences in the attentional state of the receiver. For example, in a series of observational studies, Tomasello and his colleagues have demonstrated that young chimpanzees exhibit visual signals preferentially when the putative recipient was facing towards the signaler, choosing to communicate in other modalities when the recipient was facing away (Tomasello, Call, Nagell, Olguin, & Carpenter, 1994; Tomasello, Call, Warren, Frost, Carpenter, & Nagell, 1997). Similar observational results have been reported for one lesser ape and all other great ape species, including siamangs (*Symphalangus syndactylus*: Liebal, Pika, & Tomasello, 2004b), orangutans (*Pongo pygmaeus*: Liebal, Pika, & Tomasello, 2004c), gorillas (*Gorilla gorilla*: Pika, Liebal, & Tomasello, 2003), and bonobos (*Pan paniscus*: Pika, Liebal, & Tomasello, 2005), as well as sign-language-trained chimpanzees (Bodamer & Gardner, 2002).

Experimental studies on audience effects on communication have largely validated these observational studies. Chimpanzees, for example, do not gesture in the presence of unreachable food if there is no human observer present to deliver that food (Hostetter, Cantero, & Hopkins, 2001; Leavens, Hopkins, & Bard, 1996; Leavens, Hopkins, & Thomas, 2004a). Similar results were reported for two orangutans by Call and Tomasello (1994). Thus, the influence of observer presence on the propensity to exhibit manual gestures is both strong and well-demonstrated. Experimental manipulations of the focus of visual attention of a human experimenter have also been demonstrated to influence the modality of communication by apes in captivity. We distinguish two essential findings, which parallel those found in the observational studies listed above: (a) a tendency to exhibit more visual signals when an experimenter is facing towards the signaler (Call & Tomasello, 1994; Hostetter et al., 2001; Krause & Fouts, 1997; Liebal, Pika, Call, & Tomasello, 2004a; Leavens, Hostetter, Wesley, & Hopkins, 2004b) and (b) a tendency to exhibit more auditory signals when an experimenter is facing away from the signaler (attention-getting behavior: Krause & Fouts, 1997; Leavens et al., 2004b). Thus, there is substantial agreement between observational studies of apes in communication with either other apes or with human experimenters and experimental studies of apes in interaction with human experimenters: apes adjust the modality of their communication in accordance with the attentional status of an observer. It should be noted that such findings are not universally reported by all researchers (see, e.g., Povinelli & Eddy, 1996; Theall & Povinelli, 1999); however, there are substantial procedural differences between laboratories in approaching these kinds of questions. For example, studies that found effects of audience visual attention on modality of communication have all measured the spontaneous, untrained communicative behaviors of their subjects. In contrast, studies that have failed to find these kinds of sensitivities have first subjected the apes to operant training (for example, putting a hand through a specific hole among an array of holes in a transparent screen) and then measured the response frequencies under different manipulations of observer visual attention. We suggest that it is possible that this pretraining may have interference effects on the chimpanzees' performance, perhaps changing their interpretations of task requirements. The disparity of findings in this domain deserves further study.

There are two issues with which the present study is concerned. The first concerns whether food of the same absolute size elicits the same magnitude of vocal response regardless of the size of its image at the retina. One way to assess this is to present food of similar sizes, but at different distances. If the same-sized food elicits the same vocal response irrespective of the distance at which it is presented, then this would be consistent with the idea that chimpanzees exhibit size constancy in their propensity to vocalize in the presence of food.

The other issue we explored was whether the amount of food influences the number of communicative signals in different sensory and kinetic domains, such as manual gestures, vocalizations, and visual orienting behavior. That is, does the number of different signals exhibited by chimpanzees change as a function of the amount of food presented to them? That chimpanzees display communication in different sensory domains as a function of the state of observer visual attention has received considerable recent experimental support; however, to our knowledge, there is almost no published information on the multimodal deployment of communication as a function of manipulations of the amount of available food. Leavens et al. (2005) demonstrated that chimpanzees will exhibit multiple signals much more after delivery of a less desirable food (commercial primate chow), compared to a more desirable food (a banana); thus, the quality of the food influences the number of sensory domains in which chimpanzees communicate. But we are not aware of any previous attempt to assess the effect of the amount of food on multimodal deployment of communication in chimpanzees.

In the present study we manipulated the size of desirable food and the distance of this food from each of 20 chimpanzees' cages. We had the following research questions:

a. Would chimpanzees communicate differentially in the presence of a whole banana, compared to a small piece of a banana?
b. Would chimpanzees communicate differentially as a function of the distance (or angular displacement) of food?

Method

Subjects

Subjects were 20 adolescent and adult chimpanzees (12 females, 8 males; see Table 1) housed at the Yerkes National Primate Research Center, Atlanta, Georgia, U.S.A. YNPRC is fully accredited by the National Association for Laboratory Animal Care and all relevant American Psychological Association guidelines were adhered to in the course of this study (American Psychological Association, 1992). No subject was food- or water-deprived to elicit their participation in this study. To our knowledge, none of the chimpanzees have been subject to language training.

Table 1. List of subjects by gender, age, and rearing history

Females	**Age(years)**	**Rearing history**[a]
Alice	37	Mother
Artifee	22	Nursery
Belika	36	Unknown
Cheetah	41	Unknown
Dara	10	Mother
Heppie	29	Nursery
Kengee	10	Nursery
Leslie	28	Unknown
Lucy	42	Unknown
Mega	14	Nursery
Melissa	17	Nursery
Suzanna	22	Nursery
Males		
Brian	19	Mother
Elwood	8	Nursery
Joseph	18	Nursery
Mason	14	Nursery
Merv	22	Nursery
Rufus	17	Mother
Storer	17	Mother
Winston	11	Nursery

[a] "Mother" means raised in captivity by biological mothers, "Nursery" means raised in captivity in same-aged peer cohorts, and the designation "Unknown" reflects a lack of available records, though in almost all instances these individuals were probably captured in the wild at a young age.

Procedure

The experiments were conducted during August, 1998. On any given trial, one experimenter (E1) placed either a whole banana (BIG) or an approximately 50-gram piece of a banana (SMALL) either approximately 30 cm (NEAR) or approximately 130 cm (FAR) from the left or right walls of each subject's cage (Figure 1).

Thus, there were four trial types, consisting of two trials of near food placement and two trials of far food placement; one each of the near and far trials comprised presentation of a small piece of a banana and the remaining trial in each of the far and near trials consisted of a whole banana. Each subject received one each of all four trial types in random order. Side of food placement was counterbalanced across subjects, so that each subject was presented with food in all four positions, but whether there was a BIG or SMALL food item in any of the four placements varied across subjects. Because each of the 20 subjects each received four trials, there were 80 trials conducted in total; that is, each chimpanzee experienced only one trial in each experimental condition.

After baiting, E1 departed and a second experimenter (E2) arrived, centered himself on the subject's cage at a distance of approximately 1 meter, and recorded

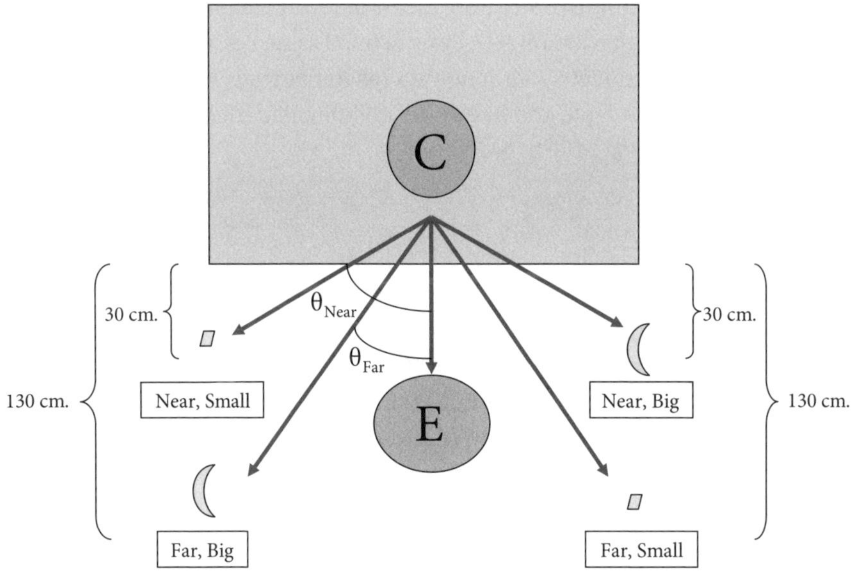

Figure 1. Schematic of experimental setup. "C" represents the subject and "E" represents the second experimenter. The position of the subject in the cage was not controlled, but in all cases: θNear > θFar. Crescents depict whole bananas and small parallelograms depict 50-gram pieces of banana. Distances are approximate; drawing is not to scale.

on a datasheet whether the subject gestured, vocalized, or exhibited gaze alternation between the food and the experimenter in a 30-second observation interval. Experimenter 2 visually engaged the subject, calling out his or her name and engaging in the kinds of unscripted verbal banter that is characteristic of interactions between humans and captive apes. At the end of each trial, the food was delivered to the subject, irrespective of whether the subject exhibited any of the behaviors of interest (i.e., a nondifferential reinforcement procedure was employed).

Behavioral coding

Manual gestures were defined as pointing with the index finger and pointing with all fingers extended (whole hand point); points were directed to the food in all cases. Food begs were defined as extended, supinated hands directed towards the experimenter, often with the hand in a "cupped" posture. A rump present is the directed orientation of the hindquarters towards a social partner. Responses defined as "other" include displays, throwing material, or spitting (see Leavens & Hopkins, 1998; Leavens et al., 2004a) for elaboration on our behavioral measures. Please note that, for present purposes, visual signals involving both manual and postural orientations are classified under the heading "manual gestures." With respect to vocalizations, chimpanzees were dichotomously classified as having either vocalized or not. Gaze alternation (GA) was defined as successive looking between the food and the experimenter. Although no reliability trials were conducted, the coding scheme is both simple and has a well-documented high reliability in other studies (e.g., Leavens et al., 2004a,b, 2005).

Analyses

Because the data were nominal to ordinal, nonparametric analyses were performed in all cases. Alpha was specified at 0.05 and all tests were two-tailed. For analyses involving dichotomous variables, Cochran's Q was used; the degrees of freedom in these analyses are the number of levels of the independent variable minus one. For ordinal data, we used Wilcoxon's approximation to the Z. Because these latter statistical analyses ignore individuals who do not change their behavior across experimental conditions, degrees of freedom are calculated only from those individuals who exhibited a change in behavior and are often much less than the total sample size of 20. For comparisons across distance, because an individual chimpanzee could exhibit up to two responses for the two FAR trials and up to two responses for the two NEAR trials, then they will be categorized during statistical analyses as having exhibited up to two responses in each of the distances involved.

For example, if a subject exhibited at least one visual signal in each of the two FAR conditions and only one visual signal in one of the two NEAR conditions, then they would be categorized as having exhibited more gestures (2–1) in the FAR condition. Note that absence of a given response was not counted as a response for these latter comparisons. Calculations were identical for all behaviors in the size of food manipulations. To assess whether the chimpanzees communicated in more sensory modalities as a function of the size of the food, the chimpanzees were categorized dichotomously as having exhibited all three of the responses (manual gestures, vocalizations, & GA) or not. In exhaustive analyses, we found no influence of either rearing history or sex of subjects on any of our dependent variables; therefore, neither rearing history nor sex will be further considered.

Results

Manual Gestures

Seventeen of the 20 chimpanzees exhibited 60 manual gestures in the course of the experiment (Table 2; only the first gesture produced in each condition was included). No systematic effect of condition on gesture type was evident (Cochran's $Q=4.00$, $df=3$, *ns*). There was a trend towards an effect of condition on vocalizations (Cochran's $Q=6.55$, $df=3$, $p=.088$) and a significant effect of condition on GA (Cochran's $Q=10.92$, $df=3$, $p=.012$; see Table 2 for frequencies). Subsequent analyses comprise separate planned comparisons within each of the manipulations: food size and food distance.

Unimodal behaviours

Neither distance nor size of the food had any influence on the chimpanzees' propensities to gesture (Wilcoxon signed ranks tests: Size, $Z(5)=-.816$, *ns* and Distance, $Z(5)=-.816$, *ns*; see Figures 2 and 3). The size of the food did not influence the subjects' propensities to exhibit GA ($Z(10)=-1.23$, *ns*), however, the size of the food did influence subjects' propensities to vocalize ($Z(11)=-2.31$, $p=.021$); the chimpanzees vocalized more frequently in the presence of the whole banana, compared to the presence of a small piece of a banana (Figure 2). Distance of the food did not influence the chimpanzees' propensities to vocalize ($Z(12)=1.00$, *ns*), however, distance did influence the subjects' propensities to exhibit GA between the food and a human observer ($Z(10)=-2.65$, $p=.008$); the chimpanzees were more likely to exhibit GA when the food was placed closer to their cages (Figure 3). Because the angular displacement between the experimenter and the food was

Table 2. Frequencies of individuals who gestured, vocalized, and exhibited gaze alternation (GA) between the food and the experimenter as a function of experimental condition ($N = 20$) and the percentage of trials in which each response occurred.

Distance:	**Near**		**Far**		
Size:	**Small**	**Big**	**Small**	**Big**	%[a]
Visual Gestures					
No Gesture	4	5	7	4	25
Indexical Point	2	1	0	1	5
Whole Hand Point	7	7	6	7	34
Food Beg	6	4	3	6	24
Rump Present	0	0	1	0	1
Other Visual Signal	1	3	3	2	11
Vocalizations					
Did not Vocalize	14	12	18	12	70
Vocalized	6	8	2	8	30
Gaze Alternation (GA)					
No GA	5	0	8	8	26
GA	15	20	12	12	74

[a] "%" means the percentage of trials in which the response was exhibited; i.e., the sum of the first four columns, divided by 80 (the total number of trials), multiplied by 100.

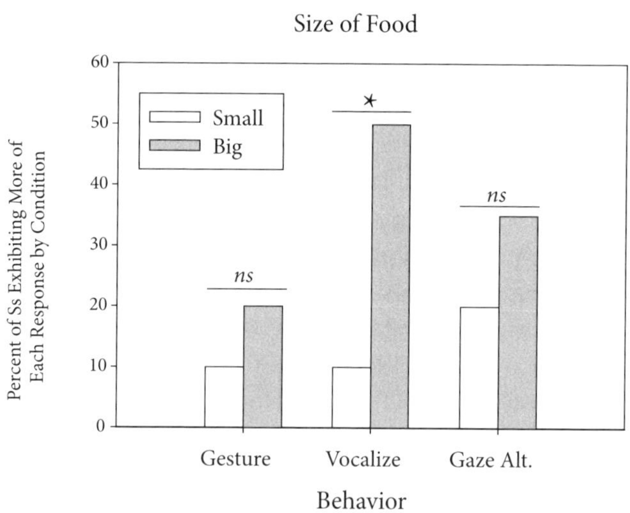

Figure 2. Influence of food size on unimodal communication by chimpanzees. The whole bananas elicited a higher propensity to vocalize than did the small pieces of banana. The asterisk denotes that $p < .05$ and "*ns*" means that the comparison was not significant.

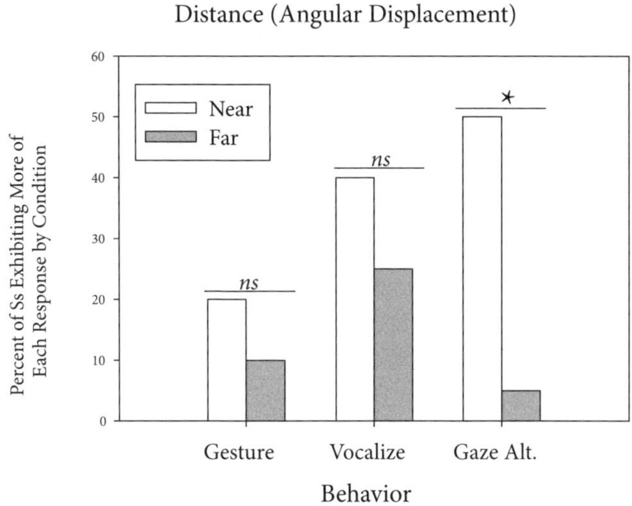

Figure 3. Influence of food distance (angular displacement) on unimodal communication by chimpanzees. Chimpanzees were more likely to alternate their gaze between the experimenter and the food when it was placed close to their cages (i.e., under conditions of higher angular displacement). The asterisk denotes that $p < .05$ and "*ns*" means that the comparison was not significant.

always larger when the food was near the cages, compared to the far placement, this means that the chimpanzees exhibited more GA when the food was placed at a greater angular displacement between the subjects and the experimenter.

Multimodal results

Subjects exhibited combinations of behaviors at higher frequencies when presented with a whole banana (gesture + vocalization + GA, $Z(6) = -2.530$, $p = .011$; see Figure 4), demonstrating an influence of the size of the food on the multimodality of signaling by the chimpanzees. In fact, not one of the 20 chimpanzees exhibited all three behaviors more in the presence of the small piece of banana, compared to the whole banana. Distance, or angular displacement, had no apparent influence on these combinations of behavior, $Z(5) = -.816$, *ns*.

Discussion

The chimpanzees in the present study exhibited size constancy in terms of their vocal output and multimodal communicative expressions in the presence of

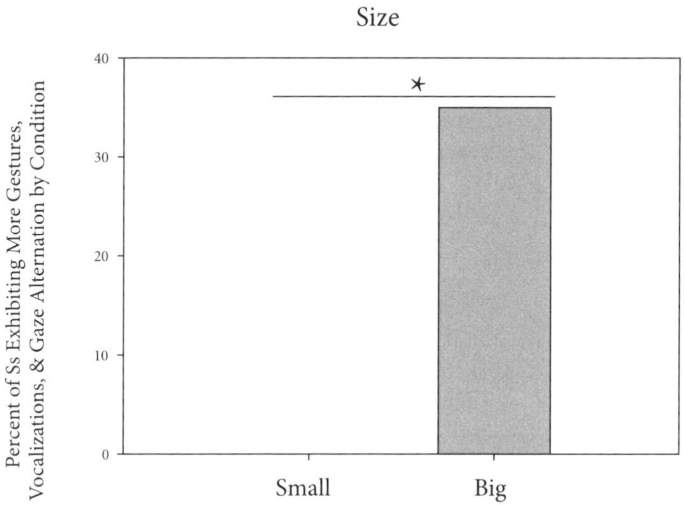

Figure 4. Influence of food size on multimodal combinations of behavior by chimpanzees. Chimpanzees deployed the full suite of manual gestures, vocalizations, and gaze alternation significantly more often in the presence of the whole banana, compared to the presence of a small piece of a banana. In fact, no chimpanzee exhibited more of this combined suite of behaviors in the presence of the small piece of banana. The asterisk denotes that $p < .05$.

unreachable food. Regardless of the distance of the food presented, chimpanzees vocalized more and exhibited multimodal signals more in the presence of the whole banana, compared to a smaller piece of a banana. The chimpanzees also exhibited more gaze alternation between food and an experimenter as the angular displacement increased between these elements (i.e., when the food was placed relatively close to the chimpanzees' cages).

In congruence with previous studies by Hauser and his colleagues (Hauser & Wrangham, 1987; Hauser et al., 1993), the chimpanzees in the present study exhibited a much higher propensity to vocalize in the presence of a whole banana, compared to the presentation of a small piece of a banana. This relationship held irrespective of the distance at which the food was presented, although there was a hint towards the possibility that a small piece of food presented at a further distance may elicit a reduced propensity to vocalize, compared to the same size of food presented at a closer distance. Specifically, only two chimpanzees vocalized in the SMALL, FAR condition, whereas six chimpanzees vocalized in the SMALL, NEAR condition (Table 2). In contrast, the chimpanzees exhibited the same high propensity to vocalize at the whole banana, irrespective of whether that banana was presented either near to or far from their cages. Hence, these chimpanzees exhibited size constancy with respect to their propensity to vocalize in the presence of a whole banana.

Additionally, irrespective of the size of the food, more peripheral placement (in the NEAR conditions — see Figures 1 & 3) elicited relatively more gaze alternation between the food and the experimenter. It is not clear from the present design whether this higher propensity to exhibit gaze alternation is due to the angular displacement, the distance of the food, or the visual access of the food to the experimenter (in the FAR conditions, the food was behind the experimenter). One intriguing possibility is that the chimpanzees discriminated the observer's visual access to the food, by exhibiting more gaze alternation when the experimenter could see both the chimpanzee and the food. The suggestion that the chimpanzees may have tactically exhibited an increased propensity to exhibit gaze alternation between the food and the experimenter when experimenter was best situated to see both the successive orienting behavior and the food is not incompatible with a number of recent studies demonstrating very sensitive deployments of communication in relation to these kinds of situational factors (e.g., de Waal, 2001; Hostetter et al., 2001; Leavens et al., 2004b; Liebal et al., 2004a; Tomasello, Hare, & Agnetta, 1999) and the question therefore warrants further study. Recent research into human communicative development has demonstrated that young children discriminate observers' visual access between about 15 to 24 months of age in their deployment of manual gestures (e.g., Franco & Gagliano, 2001; O'Neill, 1996), so the present findings suggest that it might be fruitful to explore the development of children's concomitant visual orienting behavior in response to whether or not an observer can see an indicated object

With respect to the multimodal deployment of the three dependent measures used in the present study, we found that these chimpanzees exhibited a much increased propensity to exhibit the full suite of behaviors (gestures, vocalizations, and gaze alternation) in the presence of the whole banana, compared to a 50-gram piece of a banana (Figure 4). In fact, none of the chimpanzees exhibited this particular suite of combined behaviors more in the presence of the small piece of banana, compared to the whole banana. Thus, in chimpanzees, the number of signaling elements deployed may reflect the amount of arousal or motivation to communication. This dependent measure, the display of multimodal communication, might therefore be usefully employed in future studies of the influence of food characteristics on communication by chimpanzees.

The findings that the propensity to vocalize, considered by itself, and that the propensity to exhibit multiple signals both increase in the presence of a relatively large piece of desirable food are consistent with the idea that chimpanzees are more aroused (i.e., more motivated to communicate) in the presence of a larger reward.

A reviewer commented that this might be taken as evidence against a semantic function for these signals. Although we would like to be very clear that we do not believe a semantic function for these signals is demonstrated by the present study, nevertheless we do not believe that semanticity and arousal are mutually exclusive interpretations of chimpanzee communicative signals. Two considerations are particularly relevant, here. First, Slocombe and Zuberbühler (2005) have recently reported that the agonistic screams of wild chimpanzees are acoustically distinct as a function of the transient social role of the individual involved in agonistic interactions; that is, the physical features of these chimpanzee screams changes in accordance with whether the signaler is the aggressor or the victim in an agonistic interaction. Given that the specific physical characteristics characterizing each role in agonistic interaction are, presumably, arbitrary, then despite the manifestly emotional nature of these signals, nevertheless, the arbitrariness that characterizes semantic reference may be present in the vocal repertoire of chimpanzees. The second consideration is the fairly obvious point that humans must be sufficiently aroused, or motivated, to engage in dialogue with others, and therefore to display semantic reference. Hence, in humans, semantic reference implies a minimum level of arousal, and therefore the demonstration of an influence of size of reward on the propensity to communicate cannot, by itself, unambiguously distinguish semantic functions from levels of arousal, in any species.

The present study employed a relatively small sample of chimpanzees, and our findings should, therefore, be viewed with appropriate reserve. On the other hand, despite this small sample, and consequent low-power statistical tests, we were able to replicate previous findings relating propensity to vocalize to the amount of food presented to captive chimpanzees. An ambiguity that remains in the interpretation of the present results is whether chimpanzees fail to exhibit size constancy in their motivation to communicate about relatively small food items. Recall that two chimpanzees vocalized in the FAR, SMALL condition, whereas six vocalized in the NEAR, SMALL condition; although this is not a statistically significant difference, perhaps with larger samples or more sensitive measures of vocal behavior, we would have found an interaction between the absolute size of the visible food and the distance at which it is presented. This would substantially alter our present interpretation, which is that chimpanzees exhibit size constancy in their vocal production in the presence of food presented at different distances.

In summary, the 20 chimpanzees in this study exhibited increased propensities to vocalize in the presence of a larger piece of fruit, which replicates previous findings (Hauser & Wrangham, 1987; Hauser et al., 1993). These previous findings are extended by the present finding that increased display of multimodal communication was also elicited by the larger fruit. Finally, the chimpanzees exhibited more gaze alternation between fruit and an experimenter with greater angular displacement between that experimenter and the food.

Acknowledgements

We thank two anonymous reviewers for helpful comments. This research was funded by National Institutes of Health grants RR-00165 and NS-29574.

References

American Psychological Association (1992). Ethical principles of psychologists and code of conduct. *American Psychologist, 47*, 1597–1611.

Bodamer, Mark D. & R. Allen Gardner (2002). How cross-fostered chimpanzees (*Pan troglodytes*) initiate and maintain conversations. *Journal of Comparative Psychology, 116*, 12–26.

Call, Josep & Michael Tomasello (1994). Production and comprehension of referential pointing by orangutans (*Pongo pygmaeus*). *Journal of Comparative Psychology, 108*, 307–317.

de Waal, Frans B. M. (2001, January 19). Pointing primates: Sharing knowledge … without language. *Chronicle of Higher Education*, B7-B9.

Franco, Fabia & Antonino Gagliano (2001). Toddlers' pointing when joint attention is obstructed. *First Language, 21*, 289–321.

Hauser, Marc D. (1996). *The evolution of communication*. Cambridge, MA: MIT Press.

Hauser, Marc D. & R.W. Wrangham (1987). Manipulation of food calls in captive chimpanzees. A preliminary report. *Folia Primatologica, 48*, 207–210.

Hauser, Marc D., Patricia Teixidor, L. Field, & R. Flaherty (1993). Food-elicited calls in chimpanzees: Effects of food quantity & divisibility. *Animal Behaviour, 45*, 817–819.

Hostetter, Autumn B., Monica Cantero, & William D. Hopkins (2001). Differential use of vocal and gestural communication in response to the attentional status of a human. *Journal of Comparative Psychology, 115*, 337–343.

Krause, Mark A. & Roger S. Fouts (1997). Chimpanzee (*Pan troglodytes*) pointing: Hand shapes, accuracy, and the role of eye gaze. *Journal of Comparative Psychology, 111*, 330–336.

Leavens, David A. & William D. Hopkins (1998). Intentional communication by chimpanzees: A cross-sectional study of the use of referential gestures. *Developmental Psychology, 34*, 813–822.

Leavens, David A., William D. Hopkins, & Kim A. Bard (1996). Indexical and referential pointing in chimpanzees (*Pan troglodytes*). *Journal of Comparative Psychology, 110*, 346–353.

Leavens, David A., William D. Hopkins, & Roger K. Thomas (2004a). Referential communication by chimpanzees (*Pan troglodytes*). *Journal of Comparative Psychology, 118*, 48–57.

Leavens, David A., Autumn B. Hostetter, Michael J. Wesley, & William D. Hopkins (2004b). Tactical use of unimodal and bimodal communication by chimpanzees (*Pan troglodytes*). *Animal Behaviour, 67*, 467–476.

Leavens, David A., Jamie L. Russell, & William D. Hopkins (2005). Intentionality as measured in the persistence and elaboration of communication by chimpanzees (*Pan troglodytes*). *Child Development*, *76*, 291–306.

Liebal, Katja, Simone Pika, Josep Call, & Michael Tomasello (2004a). To move or not to move: How apes adjust to the attentional state of others. *Interaction Studies*, *5*, 199–219.

Liebal, Katja, Simone Pika, & Michael Tomasello (2004b). Social communication in siamangs (*Symphalangus syndactylus*): Use of gestures and facial expressions. *Primates*, *45*, 41–57.

Liebal, Katja, Simone Pika, & Michael Tomasello (2004c). *Social communication in orangutans (Pongo pygmaeus): Use of gestures and facial expressions.* Paper presented at the Fifth International Conference on the Evolution of Language, March.

O'Neill, Daniela K. (1996). Two-year-old children's sensitivity to a parent's knowledge state when making requests. *Child Development*, *67*, 659–677.

Pika, Simone, Katja Liebal, & Michael Tomasello (2003). Gestural communication in young gorillas (*Gorilla gorilla*): Gestural repertoire, learning, and use. *American Journal of Primatology*, *60*, 95–111.

Pika, Simone, Katja Liebal, & Michael Tomasello (2005). The gestural repertoire of bonobos (*Pan paniscus*): Flexibility and use. *American Journal of Primatology*, *65*, 39–61.

Povinelli, Daniel J. & Timothy Eddy (1996). What young chimpanzees know about seeing. *Monographs of the Society for Research in Child Development*, Volume 61 (Serial no. 247). Chicago: Society for Research in Child Development.

Slocombe, Katie E. & Klaus Zuberbühler (2005). Agonistic screams in wild chimpanzees (*Pan troglodytes schweinfurthii*) vary as a function of social role. *Journal of Comparative Psychology*, *119*, 67–77.

Theall, Laura A. & Daniel J. Povinelli (1999). Do chimpanzees tailor their gestural signals to fit the attentional states of others? *Animal Cognition*, *2*, 207–214.

Tomasello, Michael, Josep Call, Katherine Nagell, Raquel Olguin, & Malinda Carpenter (1994). The learning and use of gestural signals by young chimpanzees: a trans-generational study. *Primates*, *35*, 137–154.

Tomasello, Michael, Josep Call, Jennifer Warren, G. Thomas Frost, Malinda Carpenter, & Katherine Nagell (1997). The ontogeny of chimpanzee gestural signals: A comparison across groups and generations. *Evolution of Communication*, *1*, 223–253.

Tomasello, Michael, Brian Hare, & Bryan Agnetta (1999). Chimpanzees, *Pan troglodytes*, follow eye gaze geometrically. *Animal Behaviour*, *58*, 769–777.

Requesting gestures in captive monkeys and apes

Conditioned responses or referential behaviours?

Juan-Carlos Gómez
University of St. Andrews

Captive monkeys and apes almost inevitably develop gestures to request food and objects from humans. One possibility is that these gestures are just conditioned responses without any understanding of the socio-cognitive causality underlying their efficacy. A second possibility is that they do involve some understanding of how they are (or fail to be) effective upon the behaviour of others. Observational evidence suggests that most apes and some monkeys coordinate their request gestures with joint attention behaviours — a criterion for early referential communication in human infants. However, experimental evidence about apes and monkeys' understanding of the causal role of joint attention in gestural communication is equivocal, with test pass and failure patterns that can be due to cognitive and/or motivational factors. Current evidence suggests that the gestures of apes and monkeys can neither be dismissed as simple conditioned responses nor be uncritically accepted as fully equivalent to human gestures.

Captive monkeys and apes almost inevitably develop gestures to request food and objects from humans. Many other captive animals have been anecdotally reported to engage in requesting behaviours in zoo or domestic settings. For example, Hediger (1955) reports widespread begging behaviours among zoo animals, especially among mammals, through a variety of actions — extension of the trunk (elephants), tip of the tongue (giraffes), or tail (spider monkeys). Typically primate begging involves the extension of an arm or hand in tantalising similarity to some human begging or pointing gestures.

Although early observers like Hediger thought that such begging behaviours were not typical of the natural repertoire of mammals, more recent research suggests that begging does occur naturally in a variety of species. Food transfer among different individuals has been described in many species (Stevens, 2004), including primates, in which it frequently involves "begging " behaviours (Brown,

Almond, & van Bergen, 2004). However, the exact definition of begging in this literature is unclear. Although in formal definitions, some form of gesture is implied (e.g., Brown et al., 2004), the term is frequently used in a loose sense and appears to encompass a broad variety of behaviours leading to the acquisition of food from another individual, including forceful taking and snatching (see, for example, Stevens, 2004). This lack of descriptive specification of "begging" might be due to the primary focus of such studies being functional, rather than cognitive.

In primates, food transfer involving some form of begging is most common among infants. Primate infants beg (in the broad sense) food from their parents or other adults, and there is evidence that the primary determinant of begging might be the inability to obtain the food items by themselves.

Captivity may place adult primates in a similar position of inability to obtain food or other items by themselves. Indeed captive adult primates are usually kept in cages or enclosures that prevent them from accessing outside goods. Their food and any enrichment items are brought in and given by humans. This could explain why captive primates (and non-primates) of all ages develop what otherwise would be a behaviour pattern more characteristic of infancy. (Nonetheless, begging has also been described among adult chimpanzees in situations such as meat sharing [e.g., Teleki, 1974].)

The aim of this paper is to discuss, not functional explanations of begging and requesting in primates, but whether they qualify as *referential* behaviours.

Captive primates might develop these gestures as conditioned responses without any understanding of the socio-cognitive causality underlying their efficacy. For example, chimpanzees might acquire the response of extending the arm towards a desired goal simply because they learn that this frequently results in a human giving them the desired item, without understanding how this action works as a gesture that *indicates* the human what they want. Humans may be more likely to respond to those behaviours that resemble human gestures (e.g., arm extensions), thereby shaping the animal's response into an apparent referential gesture.

To determine if non-human primates use their begging gestures with communicative intentionality, we need a set of objective criteria.

Criteria for identifying communicative gestures

In their pioneering work on the development of prelinguistic communication in human infants, Camaioni, Volterra, and Bates (1976; see also Bates, Camaioni, & Volterra, 1975) proposed a number of objective features to identify gestures as acts of intentional communication in one year-old human infants. Some of these criteria were common to identifying any action as goal-directed (e.g., variation of means and persistence until the goal is reached; see Leavens, Russell, and Hopkins [2005] for an application of such criteria to chimpanzee communication). Two criteria were, however, specifically communicative.

The first was the *schematisation of action*. Gestures are actions that are not designed to be mechanically effective. For example, pointing is not designed to grasp an object. Its efficacy depends upon its ability to make other person grasp and give the object to the communicator. However, actions that are non-mechanically effective may become schematised in a process of associative learning. Thorndike (1898) found that cats that were released from a puzzle-box upon performance of an arbitrarily chosen action (e.g., licking their paw), instead of by the accidental activation of the releasing device, tended to develop an abridged, sketched-out version of the relevant behaviour — something like a "gesture" of paw-licking.

Actions that have an effect via social causality, therefore, may have a tendency to get abbreviated as if they were gestures, but this could occur independently of any understanding of why they work. Action schematisation therefore, although necessary, may not be a sufficient criterion for determining the communicative intentionality of a gesture.

The second specific criterion proposed by Camaioni et al. (1975) — *looks at the face* of the other person by the infant — was more decisive. The reason why a gesture can be effective at all is that it is perceived by the person who has to respond to it. Checking or otherwise trying to handle the attention of the addressee (with so called "joint attention" behaviours) would be an indication that the author of a gesture understands something of the basic link between gestures and perception in communicative causality (Gómez, 1990, 1991, 2005a), and therefore uses the gesture as a referential tool. In the rest of this paper I discuss the evidence of how non-human primates use joint attention behaviours with their requesting gestures.

Gestures and joint attention in primates: Descriptive studies

Apes

Camaioni et al.'s criteria were first applied to ape communication by Tomasello et al. (1985), who found that captive chimpanzees showed gestures that fulfilled these behavioural markers of intentional communication. For example, they tended to produce visible gestures when the addressees were looking at them, which indicates they somehow checked their visual orientation.

Gómez (1990, 1991, 1992) found that a hand-reared gorilla used gestures coordinated with looks at the face of people from which she was requesting things. This longitudinal study showed that these patterns were gradually acquired. Schematised, gestural actions emerged first, and they were combined with looks at the eyes of the recipient only a few months later. Gómez suggested that looking at the eyes was a way of monitoring the attention, and not just the reaction, of the other, and therefore was suggestive of some sort of understanding of the role of attention and perception in gestural communication.

Leavens and Hopkins (1998) documented the systematic use of extended-arm gestures coordinated with attention checking in captive chimpanzees when they request food from humans. Similar behaviours have been informally reported for captive orangutans and bonobos (Gómez, 1996b) and for gibbons (Liebal, personal communication).

Monkeys

Although informal observation in captivity suggests widespread begging gestures among monkey species (Hediger, 1955), there are surprisingly few detailed reports. Blaschke and Ettlinger (1987) trained 4 rhesus monkeys to "point" ("by extending the arm and the hand") to one box that had been baited with food as a gesture to request it from a human experimenter. All the monkeys learned, but it took them an average of 428 trials — more than double the trials needed to learn a simple discrimination task (choosing one colour box over another to get the same reward), and about the same as it took a control group of monkeys to learn a spatial alternation task. Initially, the monkeys simply "reached" to the correct box, but they eventually learned to wait until the experimenter had sat in front of the boxes before extending their arm. Moreover, two of the monkeys spontaneously looked at the face of the experimenter while "pointing", and all the monkeys performed above chance in a reversed task where they had to understand the pointing gesture of a human who was trying to guide them to the box containing food (in contrast, the control group, trained only in spatial alternation, performed at chance in this comprehension task). This report therefore offers mixed evidence: on the one hand the monkeys needed a lot of training to point; on the other, some spontaneously produced joint attention behaviours, and the training was transferred to comprehension.

Povinelli, Parks, and Novak (1992) gradually trained rhesus monkeys to produce a reaching/pulling gesture in front of a baited food tray whose contents could not be seen by the human experimenter, as part of an experiment to assess whether monkeys were capable of "empathy" in a role reversal task. Although the monkeys failed the empathy task, they learned to reliably "point" to the correct food tray as a way of making the human operate on it thereby obtaining a reward. However, the gestural quality of the behaviour produced by the monkeys is not entirely clear. The authors describe under a common category of "pointing" behaviours such as "reach out of the cage in an attempt to grab the food" or "fully extending the arm out of the cage". The former would not qualify as a gesture with the criterion of action schematisation.

In a continuation of the above study, Hess, Novak, and Povinelli (1992) confirmed their negative results regarding empathy understanding with one more rhesus monkey — a sixteen-year-old female who had been hand-reared by humans during her first two years of life. This monkey needed no training to point. She had been reported to spontaneously engage in "pointing-like gestures to objects and events in the environment" with her human caretakers. The authors identify the gestures as "similar to those seen in captive chimpanzees", and remark that this spontaneous pointing was unusual in comparison to the other monkeys housed in the same laboratory, and which presumably had not been hand-reared by humans. Unfortunately, there is no detailed description of the morphology of this spontaneous pointing gesture, nor is it reported whether the pointing gestures were or not accompanied by looks at the face of the humans. However, given their failure to pass the empathy task, the authors suggest that the pointing gestures of their monkeys might be the result of conditioning rather than any complex sociocognitive understanding.

Kumashiro et al. (2002) trained two Japanese monkeys to request food from a human with a hand pointing gesture and eye contact through a process of gradual shaping of each behaviour by intensive training. One of the monkeys learned to use the pointing gesture with eye contact and even to perform gaze alternation between the food and the eyes of the human. This individual — a juvenile female — was reported to perform begging gestures with her index finger extended and to point to a TV screen in what the authors suggest could constitute an example of protodeclarative pointing. (The exact rearing history of this individual is unclear from the report).

Gómez, Lorinctz, and Perret (unpublished observations) found spontaneous arm-extended gestures to request visible pieces of food outside their cages in several members of a colony of captive rhesus macaques who had been neither hand-reared nor formally trained to point. Interestingly, some monkeys spontaneously combined their gestures with looks at the eyes of the humans from which they were begging.

Thus, two macaque species (rhesus and Japanese) seem to be capable of developing arm-extended gestures similar to those described in great apes, usually but not exclusively as a consequence of formal training. Like the apes, some of these monkeys may combine their gestures with joint attention behaviours (sometimes spontaneously, other times as a consequence of intensive training).

Behaviourally, therefore, great apes and some monkeys show what Camaioni et al. (1975) considered to be the signs of intentional referential communication. They might understand the role of attention in gestural communication.

Gestures and joint attention in primates: Experimental tests

Experimental tests, however, suggested that this might not be the case. Povinelli and Eddy (1996a, b) found that chimpanzees in a classical begging situation (trying to obtain food from a human outside their cage) were not able to choose the attentive human when given a choice to beg from someone who had the eyes open and oriented to the cage or someone who had the eyes closed or directed elsewhere. The chimpanzees did avoid humans who were with their back to them, but were unable to guide their gesturing in accordance with more subtle signs of attention and inattention. These findings appeared to be specially persuasive because in a condition where the inattentive human was looking sideways, the chimpanzees initially followed the gaze of the human and looked in the same direction, only to immediately address their request randomly. This strongly suggested that a behaviour apparently revealing an understanding of attention (gaze following) could in fact not be accompanied by any understanding of its causal role in gestural interaction.

These negative conclusions were supported by further experiments from Povinelli's lab showing that the same chimpanzees failed to understand referential gestures by humans and that any progress in their ability to select attentive over inattentive donors could be explained as associative learning of predictive cues (reviewed in Povinelli, 2000, chapter 1).

This conclusion seems to be supported by chimpanzees' difficulties to use gestures and directed gaze from humans to find hidden pieces of food in the so-called object choice paradigm, where a human tries to direct the chimpanzee to the correct choice (Call & Tomasello, 2005).

All in all, the pattern of results in strictly controlled experimental tests conducted during the 1990s suggested that chimpanzees produce their gestures without understanding how they are causally connected to the behaviour of the recipients. Their begging gestures (including the tendency to look at the eyes of humans) might be conditioned responses.

Understanding attention without gestures

However, the inability of chimpanzees to use cues of attention to guide their behaviour has been challenged by a number of recent findings. Hare et al. (2000) found that subordinate chimpanzees do take into account if dominant chimpanzees can or not see a piece of food when taking foraging decisions. The probability that they will approach a bait is higher if the dominant cannot see that particular piece of food. This suggests that chimpanzees may have some understanding of attention and perception in others, but they only use it to guide their competitive behaviour with conspecifics, not to guide their attempts at eliciting cooperative responses from humans. Their understanding of attention might remain dissociated from their ability to generate gestures.

This possibility is dramatically illustrated by a recent study by Hare and Tomasello (2004) in which chimpanzees are excellent at finding hidden food using the directional information provided by unsuccessful attempts at reaching it performed by a human, but not from similar reaching movements intended as an informative gesture to direct the chimpanzee to the food.

Therefore, the begging gestures used by captive chimpanzees may in fact not be referential gestures, as they may not be intended to direct the attention of the addressee to a target, but just to provoke a desired reaction. This would fit with the finding that apes typically use their gestures only for requesting purposes, whereas human infants also use them with the aim of calling attention with declarative and informative purposes. Some authors indeed suggest that only protodeclarative gestures are genuinely referential (see Gómez, Sarriá, & Tamarit, 1993, for a discussion).

Understanding attention with gestures

A wave of recent experiments suggests, however, that chimpanzees and other great apes may have some ability to take into account the attentional states of others when producing their requesting gestures.

Hostetter, Hopkins, and Cantero (2001), Leavens, Hopkins, and Thomas (2004), and Leavens, Hostetter, and Hopkins (2004) report that chimpanzees use vocal and manual begging gestures differentially depending upon the visual orientation of the human from which they are begging. Vocal requests are more frequent when the human is not looking at them, whereas manual gestures are preferred when the human is visually oriented to the chimpanzee.

Liebal et al. (2004) found that the four species of great apes showed signs of attention understanding in an innovative test, in which they were confronted with a human and a piece of food. The human could be oriented to or with her back to the ape. The innovation was that the apes had the opportunity of confronting the inattentive human by moving to a different part of their cage. All ape species showed a preference for moving in front of the human before producing a gesture. Chimpanzees and bonobos did so even when this implied moving away from food placed behind the human, whereas gorillas and orangutans found it more difficult to disengage from the food.

Povinelli et al. (2003) found that the direction of attention of the human addressee affected the direction in which chimpanzees performed their request gestures. When the human was actively attending to a distracter object, the chimpanzees were more likely to gesture in that direction, which suggests their gestures are not blindly guided by the target object, but also by the attentional direction of the addressee.

Apes, therefore, may find it easier to adapt their gestures to the attention direction of the human, rather than manipulating the human's attention. This fits the finding by Liebal, Call, and Tomasello (2004) that chimpanzees interacting among themselves tend to use gestures when they are in the visual range of the recipient, but do not act to call the attention of an inattentive recipient.

Kaminski, Call, and Tomasello (2004) report a complex interaction between cues of attention. Although their apes tended to show more begging responses when they were being watched by a human, the body orientation of the human was a more powerful cue than face orientation. Thus, when the human had her body oriented away from the ape, the tendency was not to respond, even if the human was actually looking at the ape over her shoulder. Conversely, when the human was oriented to the apes with body and face, these responded even if the eyes of the human were closed. The authors suggest that body orientation might be a cue signalling "willingness" to give food, and this cue would interact with broad signs of visual attention (e.g., face orientation), but not with more subtle signals (eye open vs. closed).

Gómez (1996a, 2004) reports that some chimpanzees can discriminate between lack of response and lack of attention when confronted with a human who does not immediately comply with a request, in some conditions because she is not looking at the ape, in another condition because she is just delaying the response while attending to the ape. Half of the six chimpanzees tested called the attention of the human when the lack of response was due to lack of attention, but not when it was due to a delay. Moreover, chimpanzees also called the attention of the human when she was attending to the object of their request but not to them, thereby demonstrating an understanding of the difference between attending to them and attending to an object.

Testing monkeys' use of attention and gestures

There are very few experimental studies on monkeys' understanding of attention. Many monkey species follow attention of conspecifics or humans to targets — a prerequisite for referential gestures (see review in Gómez, 2005b) — but, like chimpanzees, they find it difficult to benefit from gestural and gaze cues given by humans in object choice tests. Recent findings suggest, however, that in potentially competitive situations rhesus monkeys take into account if a human can or can not see a piece of food, as revealed by their preference to take pieces not seen by humans (Flombaum & Santos, 2005), but find it difficult to reveal a similar understanding in a cooperative situation, even when measured with an implicit response of anticipation (Lorincz et al., 2005).

We don't know if rhesus are capable of adapting their begging gestures to the attentional state of humans or call the attention of inattentive humans. The appropriate tests remain to be conducted.

In sum, current evidence offers a complex landscape of results consistent and inconsistent with apes and monkeys understanding of gestures as referential. How can we reconcile this apparently disparate set of results?

Attention to oneself and attention to other targets

One possible source of confusion is that studies not always distinguish between two different attentional components of requests — attention following or directing and attention contact or mutual attention (Gómez, 2004, 2005a). In a referential request one must manage two things: getting the attention of the addressee and directing it to the target of interest. It is this triangulation of attention between the communicator, the addressee, and the object that best characterises referential communication. A referential request must manage both attentional components, but they are separate aspects of a request, and different experiments measure one or the other (Gómez, 2005a).

Thus, the studies showing that chimpanzees or rhesus monkeys follow gaze or take into account if a potential competitor can or cannot see a piece of food show the presence of *attention following* skills — detecting the connection between an agent's attention and a target — but they tell us nothing about the *attention contact* component. A species may be capable of attention following, but not use attention contact in a request situation.

The failure of some experiments to find discrimination of the attentional availability of humans by chimpanzees may be due to a failure in their design to distinguish between attention to a third target and mutual attention between communicator and addressee. For example, in Povinelli and Eddy (1996a) the humans "oriented to" the chimpanzees avoided eye contact with them, looking instead fixedly either at a point in the Plexiglas partition or at the hole through which the chimpanzees could perform their extended-arm gestures. This eliminated crucial signs of attentional availability. Chimpanzees may have perceived that the oriented humans were attentionally as unavailable as those who were looking elsewhere or had their eyes closed. This interpretation is supported by Povinelli and Eddy (1996b), who in a different experiment found that chimpanzees preferred requesting from humans who made eye contact with them (or showed other signs of attentional orientation) over humans looking fixedly at a point as in their previous experiments.

Production and understanding

A second dimension of test variability is whether the use of attention by the ape is productive or receptive. Receptive attention following is widespread not only among primates but also other vertebrates, including some birds (see Gómez, 2005b for short review). Productively directing the attention of another individual to a target seems to be inherent in gestures oriented to a desired object. Great apes produce them readily and at least some rhesus monkeys do so as well. However, it remains to be determined if the aim of such gestures is to direct the attention or rather the action of the other upon the target (Gómez, 2005a). The looks at the eyes of the addressee spontaneously produced by apes and some monkeys might not be intended as checks of whether the gestures have succeeded in directing their attention to the target, but as checks of mutual attention, i.e., whether the addressee is or not engaged with them.

Moreover, different tests of the ability to detect and use mutual attention in requests make different demands from subjects. Some tests require to select an attentive over an inattentive person or just compare rates of gesturing between attentive and inattentive partners (e.g., Povinelli et al., 2003; Liebal, Pika, et al., 2004), whereas other tests require that the chimpanzee actively recruits the attention of an inattentive addressee (e.g., Gómez, 2004). The first type of test measures if apes can identify and use an available causal link, whereas the second requires actively repairing the missing causal link. As in the realm of tool-use, the second ability may be more complex than the former. Pulling a rake already placed behind a target is easier than having to fetch and place the rake in position. Indeed calling the attention of inattentive humans before engaging in referential requests seems to be a challenging test that only chimpanzees with extensive human-rearing experience pass (Gómez, 2004).

However, the recent innovative paradigm of Liebal et al. (2004) shows that, when given the opportunity, great apes may solve the problem of a human's inattention by changing their own relative orientation, rather than changing the humans' attentional orientation. To continue the analogy with tool-use in problem solving, this could be comparable to finding an existing roundabout route to reach a target — an ability that may be less demanding than creating such a route (e.g., by moving a box) or using a tool to bring the target within arm reach (Gómez, 2004).

Referential motives

Finally, a third dimension that is crucially relevant for requesting or begging is the contrast between competitive and cooperative motives. A request requires a unique coordination of attention following skills and mutual attention skills, but also an ability to engage in cooperative interaction. Apes and some monkeys produce gestures coordinated with joint attention that seem to assume human cooperative responses. However, as indicated by the challenging results in the object choice paradigm, they seem to have surprising problems to understand the communicative content of similar gestures produced by humans. However, in object-choice paradigms the gestures produced by the humans are not requests of food, but informative or declarative gestures intended to guide the search of the ape. The failure could therefore be due not to an inability to understand the referential nature of the gesture, but to an inability to understand the informative intention that motivates it (Hare & Tomasello, 2004; Call & Tomasello, 2005).

It remains to be determined if apes who fail to read the information contained in a declarative gesture may however read the information contained in a requesting gesture by a human. For example, in the recent Hare & Tomasello (2004) paradigm contrasting reaching actions with informative gestures, would chimpanzees realise where the reward is if the human produced a requesting gesture addressed to the chimpanzees?

Conclusions

The current state of the evidence does not allow for a simple and straightforward answer to the question highlighted in the title of this paper — are primate request gestures referential? If the criterion for reference is the spontaneous combination of gestures with joint attention behaviours (as in the early literature on human prelinguistic communication), many apes and some monkeys produce referential gestures. But if we require evidence of *understanding* the causal role of joint attention in the effectiveness of gestures, as manifest in the ability to repair faulty attentional links, the evidence is equivocal. Primates pass some tests, but fail others. All in all, evidence suggests that the requesting gestures of primates are not simple conditioned responses focused on the contingency between the gesture and the reward, but primitive referential signals based upon *some* causal understanding of the roles of attention contact and attention direction in the efficacy of the gestures. This causal understanding has some limitations, though. These limitations may be cognitive or emerge out of an interaction between the cognitive and motivational dimensions of requesting. Current evidence is insufficient to draw a complete picture of how ape and human referentiality compare, but suggests that the gestures of apes and monkeys can neither be dismissed as simple conditioned responses nor be uncritically accepted as fully equivalent to adult human referential gestures.

Acknowledgements

Sections of this paper were written as part of project REFCOM, supported by a NESTPATHFINDER grant from the European Commission, and a DGICYT grant (BSO2002-00161) from the Spanish Ministry of Science and Technology. I am grateful to Katja Liebal for insightful comments on an earlier version of this paper.

References

Bates, Elizabeth, Luigia Camaioni, & Virginia Volterra (1975). The acquisition of performatives prior to speech. *Merrill-Palmer Quarterly, 21*, 205-226.

Blaschke, M. & George Ettlinger (1987). Pointing as an act of social communication in monkeys. *Animal Behaviour, 35* (5), 1520-1523.

Brown, Gillian R., Rosamunde E.A. Almond, & Yfke van Bergen (2004). Begging, stealing, and offering: Food transfer in non-human primates. *Advances in the Study of Behavior, 34*, 265-295.

Call, Josep & Michael Tomasello (2005). What chimpanzees know about seeing revisited: An explanation of the third kind. In Naomi Eilan, Christoph Hoerl, Teresa McCormack, & Johannes Roessler (Eds.), *Joint Attention: Communication and other minds* (pp. 45-64). Oxford: Oxford University Press.

Camaioni, Luigia, Virginia Volterra, & Elizabeth Bates (1976). *La comunicazione nel primo anno di vita*. Torino: Boringuieri.

Flombaum, Jonathan I. & Laurie Santos (2005). Rhesus monkeys attribute perceptions to others. *Current Biology, 15*, 447-452.

Gómez, Juan C. (1990). The emergence of intentional communication as a problem-solving strategy in the gorilla. In Sue T. Parker & Kathleen R. Gibson (Eds.), *"Language" and intelligence in monkeys and apes: Comparative developmental perspectives* (pp. 333-355). Cambridge, Mass.: Cambridge University Press.

Gómez, Juan C. (1991). Visual behavior as a window for reading the minds of others in primates. In Andrew Whiten (Ed.), *Natural theories of mind: Evolution, development and simulation of everyday mindreading* (pp. 195-207). Oxford: Basil Blackwell.

Gómez, Juan C. (1992). El desarrollo de la comunicación intencional en el gorila. Unpublished Ph.D. Dissertation, Universidad Autónoma de Madrid.

Gómez, Juan C. (1996a). Nonhuman primate theories of (non-human primate) minds: Some issues concerning the origins of mind-reading. In Peter Carruthers & Peter K. Smith (Eds.), *Theories of theories of mind* (pp. 330-343). Cambridge, Mass.: Cambridge University Press.

Gómez, Juan C. (1996b). Ostensive behavior in the great apes: The role of eye contact. In Anne Russon, Sue Parker, & Kim Bard (Eds.), *Reaching into thought: The minds of the great apes* (pp. 131-151). Cambridge, Mass.: Cambridge University Press.

Gómez, Juan C. (2004). *Apes, monkeys, children and the growth of mind.* Cambridge, Mass: Harvard University Press.

Gómez, Juan C. (2005a). Joint Attention and the sensorimotor notion of Subject Insights from apes, normal children, and children with autism. In Naomi Eilan, Christoph Hoerl, Teresa McCormack, & Johannes Roessler (Eds.), *Joint Attention: Communication and other minds* (pp. 65-84). Oxford: Oxford University Press.

Gómez, Juan C. (2005b). Species comparative studies and cognitive development. *Trends in Cognitive Sciences, 9* (3), 118-125.

Gómez, Juan C., Encarnación Sarriá, & Javier Tamarit (1993). The comparative study of early communication and theories of mind: Ontogeny, phylogeny, and pathology. In Simon Baron-Cohen, Helen Tager-Flusberg, & Donald Cohen (Eds.), *Understanding other minds: Perspectives from autism* (pp. 397-426). Oxford: Oxford University Press.

Hare, Brian, Josep Call, B. Agnetta, & Michael Tomasello (2000). Chimpanzees know what conspecifics do and do not see. *Animal Behaviour, 59*, 771-785.

Hare, Brian & Michael Tomasello (2004). Chimpanzees are more skilful in competitive than in cooperative cognitive tasks. *Animal Behaviour, 68*, 571-581.

Hediger, Heini (1955). *The psychology and behaviour of animals in zoos and circuses.* New York: Dover.

Hess, Jo, Melinda A. Novak, & Daniel J. Povinelli (1992). 'Natural pointing' in a rhesus monkey, but no evidence of empathy. *Animal Behavior, 46*, 1023-1025.

Hostetter, Autumn B., William D. Hopkins, & M. Cantero (2001). Differential use of vocal and gestural communication by chimpanzees (*Pan troglodytes*) in response to the attentional status of a human (*Homo sapiens*). *Journal of Comparative Psychology, 115*, 337-343.

Kaminski, Juliane, Josep Call, & Michael Tomasello (2004). Body orientation and face orientation: Two factors controlling apes' begging behaviour from humans. *Animal Cognition, 7*, 216-223.

Kumashiro, Mari, Hidetoshi Ishibashi, Shoji Itakura, & Atsushi Iriki (2002). Bidirectional communication between a Japanese monkeys and a human through eye gaze and pointing. *Cahiers de Psychologie Cognitive/Current Psychology of Cognition, 21*, 3-32.

Leavens, David & William Hopkins (1998). Intentional communication by chimpanzees: A cross-sectional study of the use of referential gestures. *Developmental Psychology, 34*, 813-822.

Leavens, David, William Hopkins, & Roger K. Thomas (2004). Referential communication by chimpanzees. *Journal of Comparative Psychology, 118*, 48-57.

Leavens, David, Autumn B. Hostetter, Michael J. Wesley, & William Hopkins (2004). Tactical use of unimodal and bimodal communication by chimpanzees. *Animal Behaviour, 67*, 467-476.

Leavens, David, Jamie L. Russell, & William Hopkins (2005). Intentionality as measured in the persistence and elaboration of communication by chimpanzees (*Pan troglodytes*). *Child Development, 76*, 291–306.

Liebal, Katja, Josep Call, & Michael Tomasello (2004). Use of gesture sequences in chimpanzees (*Pan troglodytes*). *American Journal of Primatology, 64* (4), 377-396.

Liebal, Katja, Simone Pika, Josep Call, & Michael Tomasello (2004). To move or not to move: How apes adjust to the attentional state of others. *Interaction Studies, 5*, 199-219.

Lorincz, Erika, Tjeerd Jellema, Juan C. Gómez, Nick Barraclough, Den Xiao, & David Perret (2005). Do monkeys understand actions and minds of others? Studies of single cell and eye movements. In S. Dehaene, J.-R. Duhamel, M. Hauser & G. Rizzolatti (Eds.), *From monkey brain to human brain: A. Fyssen Foundation Symposium* (pp. 189-210). Cambridge, MA: The MIT Press.

Povinelli, Daniel J. (2000). *Folk physics for apes*. Oxford: Oxford University Press.

Povinelli, Daniel J. & Timothy J. Eddy (1996a). Chimpanzees: Joint visual attention. *Psychological Science, 7* (3), 129-135.

Povinelli, Daniel J. & Timothy J. Eddy (1996b). What young chimpanzees know about seeing. *Monographs of the Society for Research in Child Development, 61* (3), 1-190.

Povinelli, Daniel J., Kathleen A. Parks, & Melinda A. Novak (1992). Role reversal by rhesus monkeys, but no evidence of empathy. *Animal Behavior, 44*, 269-281.

Povinelli, Daniel J., Laura A. Theall, James E. Reaux, & Sarah Dunphy-Lelii (2003). Chimpanzees spontaneously alter the location of their gestures to match the attentional orientation of others. *Animal Behavior, 66*, 71-79.

Stevens, Jeffrey R. (2004). The selfish nature of generosity: Harassment and food in primates. *Proceedings of the Royal Society of London. Biological Sciences, 271*, 451-456.

Teleki, Geza (1974). *The predatory behavior of wild chimpanzees*. Bucknell: Bucknell University Press.

Thorndike, Edward L. (1898). Animal intelligence: An experimental study of the associative processes in animals. *Psychological Review: Series of Monograph Supplements, 2* (4), 1-109.

Tomasello, Michael, Barbara George, Ann Kruger, Michael J. Farrar, & Andrea Evans (1985). The development of gestural communication in young chimpanzees. *Journal of Human Evolution, 14*, 175-186.

Cross-fostered chimpanzees modulate signs of American Sign Language

Valerie J. Chalcraft and R. Allen Gardner
University of Nevada, Reno

Evolutionary and developmental (Evo-Devo) biologists study the interaction between genetic endowment and developmental environment (Lewontin, 2001; Robert, 2004). Cross-fostering is a powerful tool for studying Evo-Devo. Chimpanzees lived under conditions very similar to the conditions of human children with human foster families who used American Sign Language (ASL) exclusively in their presence. In this environment, cross-fostered chimpanzees acquired and used signs as human children do. Intensive analyses of extensive video records of casual conversation show that Tatu at 46–48 months directionally modulated action signs to indicate actor and instrument as human signers do. Tatu directionally modulated action signs in responses to Wh-questions such as "Who?" but directional modulations failed to appear in responses to What Demonstrative questions such as "What that?" These results confirm and extend previous results for Dar at 37–48 months. Further analyses show that Tatu also quantitatively modulated all types of signs to indicate intensity as human signers do.

Sign language studies of cross-fostered chimpanzees explore the dynamic interaction between human culture and primate biology and the intricate relationship of communicative, intellectual, and social factors in the development of individuals. Ethologists use the procedure called cross-fostering to study the interaction between genetic endowment and developmental environment when parents of one genetic stock rear the young of a different genetic stock (Stamps, 2003). Cross-fostering a chimpanzee is very different from keeping one in a home as a pet. Many people treat their pets very well and love them dearly, but pet treatment is hardly the same as child treatment. True cross-fostering — treating a chimpanzee infant like a human child in all respects, in all living arrangements, 24 hours a day every day of the year — requires a rigorous experimental regime (see R. Gardner & Gardner, 1989, for details of cross-fostering chimpanzees). In sign language studies of cross-fostered chimpanzees, cross-fosterlings developed in a nearly human home while immersed in a naturally occurring human language, American Sign Language (ASL).

Development

B. Gardner and Gardner (1980) showed how the early vocabularies of chimpanzees Moja, Pili, Tatu and Dar overlapped with the earliest vocabularies of human children as much as child vocabularies (Nelson, 1973) overlap with each other. B. Gardner & Gardner (1998) showed how semantic relations in the early phrases of Moja, Tatu, and Dar appeared in the same developmental sequence that investigators report for human children (Bloom, 1991; Bloom, Rocissano, & Hood, 1976; Braine, 1976; Leonard, 1976; Wells, 1974; De Villiers & De Villiers, 1986, pp. 50–51; Reich, 1986, p. 83). Nominative and action phrases appear first, attributives second, and experience/notice appear last in the developmental samples of children and cross-fostered chimpanzees.

Conversation

In later studies of casual conversation, Bodamer and Gardner (2002) and Jensvold and Gardner (2000) investigated how conversational probes by an interlocutor evoked contingent rejoinders. Cross-fostered chimpanzees used expansion, reiteration, and incorporation to maintain the topic of a conversation the way human adults and human children use these devices (Brinton & Fujiki, 1984; Ciocci & Baran, 1998; Garvey, 1977; Halliday & Hansen, 1976; Wilcox & Webster, 1980). Contingencies of rejoinders to probes were comparable to contingencies reported for human children (Bloom, 1991, 1993) and more comparable to older children than to younger children.

Adult cross-fosterlings integrate gaze direction and turn-taking into conversation as human speakers and signers do (Shaw, 2000). As infants, Tatu and Dar showed an immature pattern of gaze direction and turn-taking similar to immature patterns of human children.

Cheremics of ASL

The signs of a sign language (including ASL) are analogous to words in a spoken language. Just as words can be analyzed into phonemes, signs can be analyzed into cheremes, a small set of distinctive features, meaningless by themselves but combine to form meaningful morphemes, or signs that denote meaning. Stokoe (1960) distinguished cheremes that correspond to the three components of a sign:

place, configuration, and movement. The first component is the place (P) on the body or in space e.g. cheek, chest, in front of signer, etc. For example, the place for the sign glossed as WHO is the space in front of the lips. The second component is the configuration (C) e.g. fisted or open hand, which fingers are extended, how the hand is oriented toward the place. For example, the configuration for WHO is the index or hooked index extended from the fist. The third component is the type and direction of the movement (M), e.g., simple contact or rubbing, upward or downward, straight or circular movement. For example, the movement for WHO is wiggles the index or moves in circle.[1]

Modulation

Dictionaries of ASL, like dictionaries of spoken languages, show signs in citation form — the form of the reply to the question, "What is the sign for X?" The English word for X is the English *gloss* for the sign rather than the meaning (throughout this article, the English gloss for a sign is represented in capital letters). As in any living language, the visual appearance of the sign glossed as X varies from region to region and from signer to signer within a region. In addition to regional and individual variations, there are directional and quantitative modulations that are roughly constant from region to region and from signer to signer, but vary as a function of context.

Directional Modulation

Human signers indicate actor, instrument, or location with directional modulations (DM) of signs for action. DM serve as personal pronouns. Fischer and Gough (1978, p. 17) reported that almost three-fourths of the verbs in their study incorporated actor, instrument, or location.

Fant (1972, p. 75) described how DM follows the sight line:

> …an imaginary line between signer and observer, i.e. "speaker" and "listener." Whenever a sign such as SEE moves along the sight line toward the observer, the pronouns "I" and "You" are implied, thus they need not be signed.

Adult human signers usually move signs toward an adult conversational partner or object as Fant describes, but human children often place signs on the actor, instrument, or location (Ellenberger & Steyaert, 1978). As infants, the cross-fostered chimpanzees often placed signs in this immature way. For example, the citation place for QUIET is the lips of the signer but the young chimpanzees often placed QUIET on the lips of a person in contexts that called for the person to be quiet:

(1) "GRG was hooting and making other sounds, to prevent a 57 month-old Dar from falling asleep. Dar put his fist to GRG's lips and made kissing sounds. GRG asked: WHAT WANT?/ and Dar replied, QUIET/, placing the sign on GRG's lips" (GRG 5/19/81).

The "/" at the end of a sign or phrase indicates an utterance boundary.

The chimpanzees also placed signs on the instruments the adult humans used for tickling. The instruments were either objects or parts of the human other than the hand, such as the foot or mouth. As an example, KW reported in the field records:

(2) "Tickle [7 month-old] Dar with toes. Dar signs TICKLE/ (P): top of human adult's foot, (C): claw hand, (M): scratches on top of human adult's foot from ankle toward toes (KW 2/7/77).

In videotapes of casual conversation between the cross-fostered chimpanzee, Dar, and a human interlocutor, Rimpau, Gardner, and Gardner (1989) showed that Dar used DM to incorporate actor, instrument, and location into signs for action. Appropriately, Dar's answers to Who questions mostly either included name signs or pronouns or incorporated directional movement into signs for action. Equally appropriately, his answers to What Demonstrative questions (WHAT THAT?) seldom included action signs either with or without DM.

This study reports and analyzes DM in videotapes of casual conversation between the cross-fostered chimpanzee Tatu and a human interlocutor.

Quantitative Modulation

Just as human speakers indicate intensity by increasing the volume of their voices, by pausing after words or phrases, and by reiterating words, human signers indicate intensity through quantitative modulations (QM) by increasing the size and speed of signs (Friedman, 1976; Klima & Bellugi, 1979; Coulter, 1991), by holding signs in place (Friedman, 1976; Klima & Bellugi, 1979), by duplicating signs (signing a one-handed sign with two hands simultaneously) (Fant, 1972; Friedman, 1975) and by reiterating signs (Kegl & Wilbur, 1976; Rosier, 1994). R. Gardner, Gardner, and Drumm (1989) found that, like human children (Keenan, 1977), Tatu and Dar reiterated signs within an utterance to express emphasis or assent.

QM also appeared in the casual conversation of cross-fostered chimpanzees. For example, both enlarged signs and reiteration appear in the following field record:

(3) "B. Gardner signed to [25 month-old] Dar about going out to play with Tatu, and Dar agreed enthusiastically, that is, with a very large OUT OUT/" (BTG 9/4/78).

Fast or vigorous movements appeared in the following field record:

(4) "Tatu has had a few spoonfuls of carrot/cheese mixture. I hesitate with spoon. [Eight month-old] Tatu signs EAT/. P: Tatu's mouth C: loose fist, palm down M: knuckles touch P briefly (PG 8/24/76).

Movements held in place appeared in the following field record:

(5) "[Three month-old] Moja signs GO/"...thrusts out open hand toward kitchen area — continues open hand position as I walk into kitchen" (LB 2/19/73).

Duplication by using two hands simultaneously for signs that required only one hand in citation form appeared in the following field record:

(6) Tatu: PEACH/
MAG: YOU WANT MORE PEACH?/
Tatu: PEACH/ 2-handed. For intensity Tatu makes PEACH/ sign with 2 hands, one on either side of her head (MAG 8/30/78).

Sign language studies of cross-fostered chimpanzees have demonstrated continuities and comparabilities with basic aspects of human development. This is a report of a further test of the continuity and comparability of qualitative and quantitative modulation of signs in videotape records of casual conversation between cross-fostered chimpanzee Tatu and a human interlocutor.

Method

Subject

Tatu, a female chimpanzee, lived in Reno from January, 1976 to May, 1981. She was cross-fostered with Dar, a male chimpanzee, throughout this period and was cross-fostered with Moja, a female chimpanzee, from January, 1976 to December, 1979.

Cross-fostering environment

Cross-fostered infants thrived in a human environment that included human activities such as eating meals in highchairs with dishes and silverware, visiting friends, playing with toys, looking through picture books, game playing with role reversal, participating in domestic routines, and participating in toileting and napping routines. Living quarters, facilities, personal care products, food, toys, and books were all like those of human infants. Each chimpanzee had their own studio apartment or house trailer, which included a bedroom area, a kitchen area, a living area, and a bathroom. Furniture included a bed, a feeding table, an activity table with chairs, a mirror, and a dresser for storing clothes and items such as toothbrushes, hairbrushes and moisturizing lotion. The laboratory was well-stocked with the usual toys of human infants. They had various trees to climb, and fields and playrooms in which to run and explore. They often went on car rides for ice cream and went on trips to parks (R. Gardner & Gardner, 1973).

Teaching signs

The procedures that the Gardners used to teach signs were modeled after the procedures commonly used in human homes with human children. As R. Gardner and Gardner (1989, p. 15) explain:

> Most of all, we signed to each other and to the cross-fosterlings throughout the day the way human parents model speech and sign for human children. We used a very simple and repetitious register of ASL. We made frequent comments on common objects and events in short, simple redundant sentences. We amplified and expanded on their fragmentary utterances (e.g. Tatu: BLACK/ Naomi: THAT BLACK COW/). We asked known-answer questions (e.g. WHAT THAT? WHAT YOUR NAME? WHAT I DO?). We attempted to comply with requests and praised correct, well-formed utterances. All of these devices are common in human households (De Villiers & De Villiers, 1978; Moerk, 1983; Snow, 1972). Parents throughout the world seem to speak to their children as if they had very similar notions of the best way to teach languages such as English or Japanese to a young primate (Snow & Ferguson, 1977).

Field records

Throughout each day, the human members of the cross-fostering families entered observations at least once per hour, usually more frequently. Each month of field records contains 100 to 300 pages of hand-written notes. The field records originated from a long-term project in the Gardner laboratory where a team of trained graduate students and undergraduates transcribed and coded the complete field records of all the cross-fostered chimpanzees, month by month, into an electronic database.

Video records

In 15 videotapes (284 minutes), when Tatu was between 46 and 48 months of age, she interacted with one human member of her foster family, Martha Gonter

(MAG). In most of the videotapes, Tatu and MAG sit on a couch and sign about tickling, grooming, toys, food, and some personal belongings of MAG.

Transcription

Susan Nichols (SN), a member of each chimpanzee's foster family for at least 4 years with a combined total of over 15 years of cross-fostering experience, transcribed all 15 videotapes of Tatu used in the present study. Four secondary transcribers (including Beatrix T. Gardner, a principal investigator throughout all the cross-fostering in Reno) transcribed twenty-five percent of the videos for agreement. Each secondary transcriber had between five and 23 years working with the chimpanzees and their data. Agreement was calculated for gloss, utterance boundaries, and reiteration. See Rimpau et al. (1989) for transcription instructions. Agreement is reported as a percent and was calculated by the formula A/(A + D), where A was the number of items in agreement and D was the number of items that disagreed. The transcribers agreed on 84.1% of the glosses, 86.8% of the reiterations, and 89.1% of the utterance boundaries recorded in the transcript of the first transcriber. These agreement levels fall within the range accepted in studies of children (Mohay, 1982, p. 76; Siegel, 1962; Shatz & Gelman, 1973; Rondal & Defays, 1978; Snow, 1972, p. 551).

Coding

Modulations. The primary coder, Valerie Chalcraft (VC), viewed all 15 videotapes and coded Tatu's signing into the eight categories of modulations listed in Table 1. Two secondary coders categorized 28% (79 minutes) of the 15 videotapes for agreement on each modulation code. Agreement ranged from 85.1–100%.

Table 1. Modulation categories

Movement: On/Toward Object or Location
Movement: On/Toward Person
Movement: Fast or Vigorous
Movement: Held
Movement: Enlarged
Number: Duplication
Gaze at Person: Prolonged
Unable to Code

Turns. Tatu's tendency to stay on a given topic for several conversational turns serves as an independent measure of her interest in a topic of conversation. In this analysis, a conversational turn can contain one or more utterances. One partner's conversational turn ends when the next partner starts signing, whether or not the first partner stops signing. In the following illustration from videotape 8C, Tatu and MAG each have three conversational turns. Tatu's final conversational turn contains two utterances:

MAG: WHO THAT (piece of cereal) NOW?/ [i.e. "who's turn is it now?"]
Tatu: YOU SWALLOW (on MAG)/
MAG: YOU/
Tatu: ME/
MAG: WANT MORE?/
Tatu: MORE ME EAT MORE TATU/ SWEET/

Table 2. Quantitative modulation topic categories

Mask	Potty
Hat	Brush
Xylophone	Comb
Smell box (fragrance)	Oil
Key for various boxes	Toothbrush/paste
Blanket	Handkerchief/tissue
Horse	Self-groom hurts
Cow	Bib
Doll	Sandwiches
Bear	Carrots
Dinosaur	Milk
Human adult's wristwatch	Dry cereal in bag or box
Human adult's eyeglasses	Grapes
Flower	Gum
Human adult's shoe	Candy in bag or box
Phonograph	Crackers
Metal clamp	Cookies
Human adult makes animal sounds	Popcorn
Human adult laughs and cries	Nuts
Furniture (chair, bed, phonograph), include sitting	Sodapop
Human adult's hurt (wound)	Juice
Human adult's and chimpanzee's clothes	Human adult's coffee
Picture book	Apple
Label human adult/chimpanzee	Orange
End session/attempt to leave	Unspecified present food
Wash face/wash hands/soap	Unable to code

Table 2 shows the 51 topics about which Tatu and the human interlocutor signed in the 15 videotapes. Topics include different types of food (e.g., grapes, gum, cookies), grooming tools (e.g., brush, comb, oil), toys (e.g., masks, dolls), games (e.g., chase, tickle), clothes (e.g., hats, shoes, glasses), and interactions with humans (e.g., laughing, animal sounds). From the transcripts of all 15 videotapes, the primary coder coded each of Tatu's turns into one of 51 topic codes corresponding to the activities listed in Table 2. The topic of Tatu's signs rather than the topic of MAG's signs determined the code, even though MAG sometimes attempted to shift topics while Tatu remained on the original topic. One secondary coder categorized 66% (187 minutes) of the 15 videotapes for agreement on each topic code. Agreement ranged from 85.0%–100%.

Results

Directional modulation

The present study of Tatu replicated the findings that Rimpau et al. (1989) reported on directional modulations for Dar. Like human signers, Tatu and Dar distributed DM across sign categories and Wh-question contexts.

Directional modulation and sign category

Table 3 lists Tatu's signs in the 15 video taped samples categorized as names/pronouns, locatives, markers, common nouns, nouns/verbs, verbs, and modifiers. The categories agree with Rimpau et al. (1989), B. Gardner and Gardner (1975), R. Gardner, Van Cantfort, and Gardner (1992), Brown (1968), and Ervin-Tripp (1970) (see Table 2 in B. Gardner and Gardner for a sample of replies classified into general categories).

Table 3 shows the total number of different signs (types), the number of occurrences (tokens) of each sign as well as the number of sign tokens with DM. If Tatu used DM to incorporate reference to actor, instrument, or location into signs for action then her DM should occur more frequently in the verb and noun/verb categories and less frequently in other categories. Table 4 shows that this is the case.

Table 4 shows the total number of different signs (types) and number of types with DM for each general category. Twelve out of 15 signs (80.0%) classified as noun/verbs and 6 out of 14 signs (42.9%) classified as verbs occurred with DM. Added together, 62.1% of signs in these two categories occur with DM. In contrast, the locative category has one in four (25.0%) and the modifier category had one in five signs (20.0%) that occurred with DM. Twelve of 43 common noun signs (27.9%) occurred with DM, 2 of 10 marker signs (20.0%) occurred with DM,

Table 3. Signs reported in Tatu's videotapes classified into general categories. (with directional modulations)

NAMES/PRONOUNS		COMMON NOUNS		NOUNS/VERBS		VERBS	
N	Item	N	Item	N	Item	N	Item
1	DAR	39	APPLE(6)	40	BED	88	BITE(48)
55	MAG	7	BABY	41	BRUSH(6)	1	CATCH
161	ME	1	BANANA	43	CLEAN(2)	26	CRY(1)
13	ML	113	BERRY	121	COMB(6)	12	GROOM(5)
6	RAG	3	BIB	255	DRINK(19)	3	GO
405	TATU	10	BIRD	228	EAT(29)	2	HUG
135	YOU	6	BLANKET	16	HEAR(4)	2	KNOW
LOCATIVES		6	BOY	61	HANDKERCHIEF(8)	18	LAUGH(3)
27	HERE	12	CARROT	11	LISTEN(1)	13	OPEN
10	HOME (3)	27	CAT	47	OIL(5)	5	PEN/WRITE
8	OUT	44	CEREAL	34	PEEKABOO(1)	7	QUIET
469	THAT	47	COFFEE(8)	11	POTTY	68	SWALLOW(23)
MARKERS		34	COOKIE	30	SEE(4)	1	THINK
6	CAN'T	1	CORN	27	SMELL	47	TICKLE(43)
5	DIRTY	79	COW	46	TOOTHBRUSH(2)	MODIFIERS	
48	FINISH(2)	17	CRACKER(1)			52	BLACK(1)
28	GIMME	2	CUP			32	GOOD
23	HURRY	15	DOG			107	MINE
91	MORE(1)	7	EARRING			10	RED
8	NO	31	FLOWER(4)			24	YOURS
99	PLEASE	7	GLASS				
2	WHO	130	GRAPES(19)				
1	YES	38	GUM(7)				
		77	HAT(25)				
		18	HORSE				
		5	HURT				
		5	KEY				
		5	LIPSTICK				
		16	MEAT				
		13	MEDICINE				
		14	MILK				
		14	NUT				
		17	ORANGE				
		2	PEACH				
		1	ROCK				
		7	SANDWICH				
		15	SHIRT(5)				
		19	SHOE				
		86	SODAPOP(8)				
		67	SWEET				
		18	TOOTHPASTE(1)				
		40	WIPER(1)				
		13	WRISTWATCH(7)				

N = Number of reports in taped samples
Number in ( ) is total number of times the sign was reported with a modulation

Table 4. Signs with directional modulations. Number Of Signs (TYPES) Reported By Category

CATEGORY	FREQUENCY		PERCENT
	TOTAL TYPES	MODULATED TYPES	MODULATED SIGNS
NAMES/PRONOUNS	7	0	0.0
LOCATIVES	4	1	25.0
MARKERS	10	2	20.0
MODIFIERS	5	1	20.0
COMMON NOUNS	43	12	27.9
NOUN/VERBS	15	12	80.0
VERBS	14	6	42.9
TOTALS	98	35	35.7

and none of the seven names/pronouns in this sample occurred with DM. Added together, only 23.2% of signs from outside the verb and noun/verb categories occurred with DM. Eighteen out of 35 (51.4%) of Tatu's total sign types with DM were verbs or noun/verbs. This is similar to Rimpau et al's (1989) report that 52.4% of Dar's sign types with DM were verbs or noun verbs.

Context analysis of directional modulations

Of the many types of questions that MAG asked Tatu, there was one category of question that should have excluded reference to actor, instrument, or location and that was the What Demonstrative type (cf. B. Gardner & Gardner, 1975; Rimpau et al., 1989; R. Gardner et al., 1992). The What Demonstrative includes questions such as WHAT THIS? and NAME THIS? For example, an appropriate reply to WHAT THAT? (indicating brush) would be BRUSH.

Verb signs are inappropriate replies to What Demonstrative questions. Therefore, modulations or additional signs to indicate actor, instrument, or location should be unnecessary or inappropriate.

Appropriate replies to other Wh-Questions (such as WHO, WHICH, and WHERE) could contain signs from any of the general categories listed in Table 3. For example, an appropriate reply to WHO BRUSH?/ would be YOU BRUSH/. Therefore Tatu could indicate actor by signing a verb or noun/verb in citation form with an additional sign such as a personal pronoun. Alternatively, Tatu could indicate actor by directionally modulating a verb or noun/verb.

Citation form only. The following analysis examined Tatu's answers to What Demonstrative vs. other Wh-questions that contained only citation forms of all the signs types in the verb and noun/verb categories in Table 2. In reply to What

Demonstrative questions, a sign in citation form with an additional sign to indicate actor, instrument, or location, is unnecessary. For example, MAG: WHAT THAT?/ (brush) Tatu: BRUSH/. In reply to other Wh-questions, a sign in citation form is not always sufficient and may be accompanied by a second sign to indicate actor, instrument, or location. For example, MAG: WHERE BRUSH?/ (brush) Tatu: BRUSH THERE/.

Table 5 shows the distribution of the 408 citation forms that occurred in replies to the questions of MAG (chi-square = 20.36, d.f. = 1, p<0.0001). These results show that in replies to questions other than Wh-demonstrative questions, 88.4% of Tatu's signs that referenced actor, instrument, or location appeared with verbs and noun/verbs in citation form. Conversely, in replies to What Demonstrative questions, only 11.5% of Tatu's signs that referenced actor, instrument, or location appeared with verbs and noun/verbs in citation form.

The present study used a larger pool of data but confirmed the results of Rimpau et al. (1989). Using only Dar's 13 most common verb and noun/verb sign types, Rimpau et al. found that Dar added a sign to indicate actor, instrument, or location in 91.1% of replies to other Wh-Questions and in 9.9% of replies to What Demonstrative questions.

What Demonstrative questions vs. Who questions. This analysis used all the noun, noun/verb, and verb sign types in Tatu's videotapes (see Table 2) that occurred with DM toward person or in citation form in response to What Demonstrative questions and Who questions. Table 6 shows the distribution of DM toward person and citation forms according to question type in the videotapes. Most of the replies (90.9%) with a DM toward person occurred in reply to Who questions. This is where one would expect a proper noun or personal pronoun. Citation forms occurred in reply to both types of questions. This is an expected outcome, since citation forms can be combined with personal pronouns. The significant chi-square (32.07 d.f. =1, p < 0.0001) shows that questions exerted control over the form of signs in replies.

The present study used a larger pool of data but confirmed the results of Rimpau et al. (1989). Using only Dar's nine most commonly modulated noun, noun/

Table 5. Additional Signs. Indicating Actor, Instrument, or Location in Replies. Containing Verbs and Noun/Verbs in Citation Form

	Present	Absent	Totals
What Demonstrative questions	13	102	115
Other questions	100	193	293
Totals	113	295	408

Chi-square = 20.36, d.f. = 1, p<0.0001

Table 6. Citation Vs. On/Near Person According to Question Type in Replies Containing Common Nouns, Noun/Verbs, and Verbs

	What Demonstrative	Who/Whose	Totals
Citation	46	219	365
On Person	9	90	99
Totals	155	309	464

Chi-square = 32.07, d.f. = 1, $p<0.0001$

verb, and verb sign types, Rimpau et al. found that 63% of Dar's DM toward person occurred in reply to Who questions.

Quantitative modulation

Rimpau et al. (1989) only studied DM. The present study includes QM. Like human signers, Tatu used a variety of QM in all sign categories. Tatu's QM correlated significantly with an independent measure of intensity.

Distribution of Quantitative Modulation

Table 7 shows that Tatu increased the speed and size of signs, held signs, reiterated signs and duplicated signs. Table 7 also shows how Tatu signed some sign tokens with single instances of QM and some with multiple instances of QM. For example, Tatu signed 273 sign tokens with increased speed alone, signed 47 sign tokens with both increased speed and increased size and signed 123 sign tokens with increased speed, increased size and reiteration.

Table 8 lists the sign types reported in the 15 videotapes of Tatu interacting with MAG and categorizes them as names/pronouns, locatives, markers, common

Table 7. Distribution of Quantitative Modulations. per Sign Token Within Conversational Turn

Single Modulations		**Two Combined**		**Three Combined**	
Type	Frequ.	Type	Frequ.	Type	Frequ.
NR (Reiterated)	476	NR + ME	343	NR + ME + MB	123
ME (Fast/Vigorous)	273	NR + MB	104	NR + ME + ML	8
ML (Held)	244	NR + ML	80	NR + NB + ML	1
MB (Enlarged)	29	MB + ME	47	NR + MB + ML	1
NB (Duplicated)	1	ME + ML	12		
		NB + NR	3		
		MB + ML	3		
Total Single Modulations	**1023**		**Total Combinations**		**725**

Table 8. Signs Reported In Tatu's Videotapes Classified Into General Categories. With quantitative modulations.

NAMES/PRONOUNS		COMMON NOUN		NOUNS/VERBS		VERBS	
N	Item	N	Item	N	Item	N	Item
1	DAR(1)	39	APPLE(22)	40	BED(17)	88	BITE(40)
55	MAG(16)	7	BABY(2)	41	BRUSH(20)	1	CATCH(1)
161	ME(74)	1	BANANA	43	CLEAN(23)	26	CRY(17)
13	ML(2)	113	BERRY(49)	121	COMB(43)	12	GROOM(4)
6	RAG	3	BIB	255	DRINK(101)	3	GO(1)
405	TATU(212)	10	BIRD(3)	228	EAT(70)	2	HUG(2)
135	YOU(24)	6	BLANKET(2)	16	HEAR(4)	2	KNOW
LOCATIVES		6	BOY(2)	61	HANDKERCHIEF(26)	18	LAUGH(9)
27	HERE(15)	12	CARROT(3)	11	LISTEN(7)	13	OPEN(8)
10	HOME(2)	26	CAT(11)	47	OIL(18)	5	PEN/WRITE(1)
8	OUT(4)	44	CEREAL(31)	34	PEEKABOO(24)	7	QUIET(2)
469	THAT(209)	47	COFFEE(20)	11	POTTY(4)	68	SWALLOW(19)
MARKERS		34	COOKIE (13)	30	SEE(7)	1	THINK
6	CAN'T(2)	1	CORN	27	SMELL(7)	47	TICKLE(12)
5	DIRTY(3)	79	COW(64)	46	TOOTHBRUSH(16)	1	WASH(1)
48	FINISH(3)	17	CRACKER(7)			MODIFIERS	
28	GIMME(1)	2	CUP(1)			52	BLACK(27)
23	HURRY(7)	15	DOG(12)			32	GOOD(10)
91	MORE(31)	8	EARRING(5)			107	MINE(52)
8	NO(3)	31	FLOWER(18)			10	RED(5)
99	PLEASE(47)	7	GLASS(5)			24	YOURS(5)
2	WHO	130	GRAPE(66)				
1	YES	38	GUM(12)				
		77	HAT(38)				
		18	HORSE(6)				
		5	HURT(3)				
		5	KEY(1)				
		5	LIPSTICK(3)				
		16	MEAT(3)				
		13	MEDICINE(5)				
		14	MILK(5)				
		14	NUT(5)				
		17	ORANGE(2)				
		2	PEACH(1)				
		1	ROCK(1)				
		7	SANDWICH(1)				
		15	SHIRT(6)				
		19	SHOE(10)				
		86	SODAPOP(40)				
		67	SWEET(28)				
		118	TOOTHPASTE(4)				
		40	WIPER(11)				
		13	WRISTWATCH(3)				

N = Number of reports in taped samples
Number in ( ) is total number of modulations for each sign type

nouns, nouns/verbs, verbs, and modifiers. Table 8 shows the total number of types, the number of tokens of each sign, as well as the number of sign tokens with QM.

The following analysis quantifies Tatu's distribution of QM to determine if QM occurs across sign categories or, like DM, occurs mostly in specific categories or specific signs. Table 9 shows the total number of sign types and number of sign types with QM. Table 9 shows that across the categories, Tatu modulated at least one dimension of QM in 91.8% of signs and this distribution ranges between 85.7% and 100% throughout the categories. Therefore, Tatu quantitatively modulated signs in all categories and in roughly equal proportion.

Quantitative modulation and intensity of interest

Tatu's freedom to continue on a conversational topic or to break off at will offers a measure of her interest in each conversational topic. Thus, MAG could start a topic of conversation, by bringing with her items such as brushes or hats that could serve as conversational openers, and by shifting topics when she (MAG) tired of a topic. Even so, Tatu determined the number of consecutive conversational turns on any particular topic. Tatu could fail to respond to any of MAG's initial conversational gambits, breaking the thread of conversation immediately, or she could respond and continue to respond as long as MAG stayed on that topic. Tatu could even stay on a topic after MAG attempted to shift to something new. Consequently, the number of consecutive conversational turns on a particular topic measures Tatu's interest in that topic. For example, summed across all 15 video taped sessions, there were 54 total conversational turns coded with the topic code "grapes", and 51 of these were consecutive turns. There were 29 instances of QM during the consecutive conversational turns coded "grapes". Each topic code received the

Table 9. Signs with quantitative modulations. Number Of Signs (TYPES) Reported By Category

	FREQUENCY		PERCENT
CATEGORY	TOTAL TYPES	MODULATED TYPES	MODULATED SIGNS
NAMES/PRONOUNS	7	6	85.7
LOCATIVES	4	4	100.0
MARKERS	10	8	80.0
MODIFIERS	5	5	100.0
COMMON NOUNS	43	40	93.0
NOUN/VERBS	15	15	100.0
VERBS	14	12	85.7
TOTALS	98	90	91.8

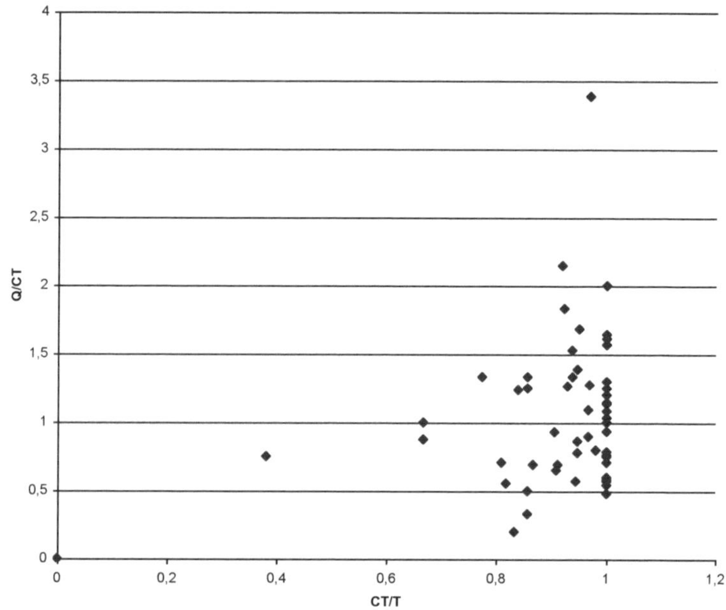

Figure 1. Conversational interest vs. Quantitative Modulations where CT/T= the ratio of consecutive turns in a topic to all turns in that topic and Q/CT=the ratio of QM in consecutive turns in a topic to the total number of consecutive turns in that topic.

following scores: T (total number of conversational turns on that topic), CT (total number of consecutive conversational turns on that topic), and Q (total number of QM in all consecutive turns in that topic). Thus, CT/T is the ratio of consecutive turns on a topic to all turns on that topic and measures Tatu's conversational interest. And, Q/CT is the ratio of QM in consecutive turns in a topic to the total number of consecutive turns in that topic and measures intensity during consecutive turns. Figure 1 shows that Q/CT (the measure of QM) correlated $r = 0.45$ (Pearson Product Moment) with CT/T (the measure of conversational interest) ($p<.001$). That is, QM correlated positively with an independent measure of conversational interest.

Discussion

Findings

As Rimpau et al. (1989) found for Dar, Tatu directionally modulated verbs and noun/verbs more than other types of signs. That Tatu incorporated actor, instrument, or location into 60% of verb and noun/verb types is also similar to the report of Fischer and Gough (1978, p. 17) who reported that adult, native signers moved almost three-fourths of their verbs toward persons or objects. Directionally modulating verbs and noun/verbs to indicate actor, instrument, or location is like incorporating pronouns into verbs and thus is more appropriate than directionally modulating other types of signs to indicate actor, instrument, or location.

As Rimpau et al. found for Dar, Tatu signed in citation form with more additional signs to indicate pronouns in replies to questions other than What Demonstrative questions than in replies to What Demonstrative questions. Further, like Dar, Tatu used DM more in reply to Who questions than in reply to What Demonstrative questions. Directionally modulating signs to indicate actor, instrument, or location and using citation form with additional signs to indicate actor, instrument, or location is more appropriate in replies to questions other than What Demonstrative questions than in replies to What Demonstrative questions.

Tatu signed with a variety of quantitative modulations (QM). Like human signers, she increased the speed and size of signs, she held signs, she reiterated signs, and she duplicated signs. Analysis of QM showed that Tatu used QM in all categories of signs at a roughly equal rate.

Investigators of human sign language often attribute quantitative modulation to intensity (Fant, 1972; Kegl & Wilbur, 1976; Klima & Bellugi, 1979; Livingston, 1983; Rosier, 1994; Shepherd, 1990, 1994). Investigators can confirm their interpretations by asking human adults to express intensity or by asking them what they mean to express when they use quantitative modulation in particular cases. Credible interviews and self-reports of human children and young chimpanzees are probably impossible to obtain. In spite of this problem, several researchers (Keenan, 1977; Livingston, 1983; Shepherd, 1990, 1994) have asserted that human children use QM, such as reiteration, to express intensity or stress. Their familiarity with deaf children and sign language lends a certain credibility to otherwise unsupported inferences on the part of these writers.

The free conversational style of Tatu's verbal interaction with MAG in the present study offers an independent measure of interest in different conversational topics. Tatu was free to use QM as much or as little as she liked in any conversational turn. The fact that QM and consecutive conversational turns on a topic correlated experimentally confirms the hypothesis that, like human signed conversation, QM indicates interest and intensity in the conversation of chimpanzees. The present study on QM builds on the findings of R. Gardner, Gardner, and Drumm (1989) who found that both Tatu and Dar reiterated more in their signed replies to positive announcements than in their signed replies to negative and neutral announcements.

Implications

Theorists such as Chomsky (1979) and Pinker (1994, pp. 332–369) treat human language as dependent on a special gene only found in the human genome virtually independent of developmental environment. As Moore (1973) precisely expressed the early Chomskian view point on human language acquisition:

> Children all over the world learn to speak their native language at approximately the same time — 3 to 4 years of age. Within a relatively brief period, the child appears to learn a complicated and abstract system of rules — a system which enables him to produce and understand an infinite variety of utterances....without teaching or training, [human children] readily acquire their native language at about the same time — regardless of just about any variable one cares to look at, short of deafness or severe retardation. (Moore, 1973, p. 4)

Developments in modern genetics indicate that there are just too few genes in the human genome to support such a simple-minded view. Any phenomenon as complex as human language must be a product of complex interactions between many genes. Moreover, modern studies of still more complex interactions between genetic endowment and developmental environment emphasize the failure of such modular views of complex phenomena.

The success of Project Washoe stimulated synthetic studies of language-like tasks with a variety of nonhuman animals (e.g. Greenfield & Savage-Rumbaugh, 1990; Herman, Kuczaj, & Holder, 1993; Matsuzawa, 1985; Pepperberg, 1992; Premack, 1971; Rumbaugh, 1977; Gisiner & Schusterman, 1992; Terrace, Petitto, Sanders, & Bever, 1979; Savage-Rumbaugh, 1984), each study with its own theoretically defined components and highly specific tasks. In contrast with the arbitrary fixed units of these "ape language" and "animal language" projects, the signs and words of human interactive conversation lend themselves to meaningful modulation. Rimpau et al.'s (1989) analysis of informal conversations with Dar and the present analyses of informal conversations with Tatu demonstrate that cross-fostered chimpanzees modulate their signs to indicate actor, instrument, and location with directional modulations that resemble directional modulations observed in human signers. In addition, the present analyses of Tatu show that she quantitatively modulated her signs as human signers do to express intensity. Further, the present analyses show that Tatu's quantitative modulations correlated strongly with an independent measure of her interest in different conversational topics. Without training beyond treating them as far as possible like human children, cross-fostered chimpanzees developed characteristically human modulations of signs in a naturally occurring human language. Sign language studies of cross-fostered chimpanzees emphasize the powerful contribution of developmental environment and the integration of semantics, structure, and pragmatics into coherent and contingent conversation.

Ethological studies in Reno and Ellensburg have looked for the give and take quality of spontaneous conversation. Synthetic studies test for hypothetical aptitudes and capacities under constrained testing conditions. Human speakers and signers modulate the forms of words and signs in rule-governed ways that depend on the verbal and nonverbal context of every day conversation whereas the symbols of synthetic studies have fixed referents and fixed forms based on Aristotelian traditions of language and thought. Synthetic studies ignore the fact that terms in natural languages mostly represent fuzzy categories that are faithful to the overlapping categories of the natural world (Zadeh & Kacprzyk, 1992).

Truly discontinuous phenomena must be rare in nature. Historically, the great discontinuities have proved to be conceptual barriers rather than rifts in the fabric of the natural world. It seems unlikely that a phenomenon as rich as language could be based on an isolated, unitary biological trait. It is more reasonable to suppose that language is the result of a complex of interacting traits running through all aspects of human intelligence. Following the same line of reasoning we would argue that, like other significant biological phenomena, the general principles that govern human intelligence are related to the general principles that govern the intelligence of all animals. This search for general biological principles of intelligence led to sign language studies of cross-fostered chimpanzees (B. Gardner & Gardner, 1998; R. Gardner & Gardner, 1998).

Note

1. The P, C, and M of the signs of the cross-fostered chimpanzees are listed in Table 3.2 of B. Gardner, Gardner, and Nichols (1989). Table 3.2 consists of the PCMs for 223 signs that met the stringent criteria for reliability in the cross-fostering laboratory. B. Gardner et al. contains eight pages of legend and comments for Table 3.2, which include discussion of criteria for sign reliability. In the present study, all of the chimpanzees' signs in the videotaped data are listed in Table 3.2, except for the signs, THINK and KNOW, which had not yet become reliable signs in the chimpanzees' vocabulary.

References

Bloom, Lois, Lorraine Rocissano, & Lois Hood (1976). Adult-child discourse: Developmental interaction between information processing and linguistic knowledge. *Cognitive Psychology*, *8*, 521–552.

Bloom, Lois (1991). *Language development from two to three*. New York: Cambridge University Press.

Bloom, Lois (1993). *The transition from infancy to language: Acquiring the power of expression*. New York: Cambridge University Press.

Bodamer, Mark D. & R. Allen Gardner (2002). How Cross-fostered chimpanzees *(Pan troglodytes)* initiate and maintain conversations. *Journal of Comparative Psychology, 116*, 12–26.

Braine, Martin D. (1976). Children's first word combinations. *Monographs of the Society for Research in Child Development, 12* (1, Serial No. 164).

Brinton, Bonnie & Martin Fujiki (1984). Development of topic manipulation skills in discourse. *Journal of Speech and Hearing Research, 27*, 350–358.

Brown, Roger (1968). The development of Wh-questions in child speech. *Journal of Verbal Learning and Verbal Behavior, 7*, 279–290.

Chomsky, Noam (1979). Human language and other semiotic systems. *Semiotica, 25*, 31–44.

Ciocci, Sandra R. & Jane A. Baran (1998). The use of conversational repair strategies by children who are deaf. *American Annals of the Deaf, 143*, 235–245.

Coulter, Geoffrey R. (1991). Intense stress in ASL. In Susan D. Fischer & Patricia Siple (Eds.), *Theoretical Issues in Sign Language Research*. (pp. 109–125). Chicago: University of Chicago Press.

De Villiers, Jill G. & Peter A. De Villiers (1978). *Language acquisition*. Cambridge, MA: Harvard University Press.

De Villiers, Jill G. & Peter A. De Villiers (1986). The acquisition of English. In Dan I. Slobin (Ed.), *The crosslinguistic study of language acquisition* (Vol. 1). Hillsdale, NJ: Lawrence Erlbaum.

Ellenberger, Ruth & Marcia Steyaert (1978). A child's representation of action in ASL. In Patricia Siple (Ed.), *Understanding language through sign language research* (pp. 261–269). New York: Academic Press.

Ervin-Tripp, Susan (1970). Discourse agreement: How children answer questions. In John R. Hayes (Ed.), *Cognition and the development of language* (pp. 79–107). New York: John Wiley & Sons.

Fant, Louis (1972). *Ameslan: An introduction to American Sign Language*. Northridge CA: Joyce Motion Picture Co.

Fischer, Susan & Bonnie Gough (1978). Verbs in American Sign Language. *Sign Language Studies, 18*, 17–48.

Friedman, Lynn A. (1975). Space, time, and person reference in American Sign Language. *Language, 51*, 940–961.

Friedman, Lynn A. (1976). Phonology of a soundless language: Phonological structure of American Sign Language. Unpublished doctoral dissertation. University of California, Berkeley.

Gardner, Beatrix T. & R. Allen Gardner (1975). Evidence for sentence constituents in the early utterances of child and chimpanzee. *Journal of Experimental Psychology: General, 104*, 244–267.

Gardner, Beatrix T. & R. Allen Gardner (1980). Two comparative psychologists look at language acquisition. In Keith E. Nelson (Ed.), *Children's language* (Vol. 2, pp. 331–369). New York: Gardner Press.

Gardner, Beatrix T., R. Allen Gardner, & Susan G. Nichols (1989). The shapes and uses of signs in a cross-fostering laboratory. In R. Allen Gardner, Beatrix T. Gardner, & Thomas. E. van

Cantfort (Eds.), *Teaching sign language to chimpanzees* (pp. 55–180). Albany, NY: SUNY Press.

Gardner, Beatrix T. & R. Allen Gardner (1998). Development of phrases in the early utterances of children and cross-fostered chimpanzees. *Human Evolution, 13*, 161–188.

Gardner, R. Allen & Beatrix T. Gardner (1973). *Teaching Sign Language to the Chimpanzee, Washoe* (16 mm sound film). State College, PA: Psychological Cinema Register.

Gardner, R. Allen & Beatrix T. Gardner (1998). Ethological study of early language. *Human Evolution, 13*, 189–207.

Gardner, R. Allen & Beatrix T. Gardner (1989). A cross-fostering laboratory. In R. Allen Gardner, Beatrix. T. Gardner, & Thomas E. Van Cantfort (Eds.), *Teaching sign language to chimpanzees,* (pp. 1–28). Albany, NY: SUNY Press.

Gardner, R. Allen, Beatrix T. Gardner, & Patrick Drumm (1989). Voiced and signed replies of cross-fostered chimpanzees. In R. Allen Gardner, Beatrix T. Gardner, & Thomas E. Van Cantfort (Eds.), *Teaching sign language to chimpanzees* (pp. 29–54). Albany, NY: SUNY Press.

Gardner, R. Allen, Thomas E. van Cantfort, & Beatrix T. Gardner (1992). Categorical replies to categorical questions by cross-fostered chimpanzees. *American Journal of Psychology, 105*, 27–57.

Garvey, Catherine (1977). Contingent queries and their relations in discourse. In Elinor Ochs & Bambi Schieffelin (Eds.), *Developmental pragmatics* (pp. 363–372). New York: Academic Press.

Gisiner, Robert & Robert J. Schusterman (1992). Sequence, syntax, and semantics: Responses of a language-trained Sea Lion to novel sign combinations. *Journal of Comparative Psychology, 106*, 78–91.

Greenfield, Patricia M. & E. Sue Savage-Rumbaugh (1990). Grammatical combination in *Pan paniscus:* Processes of learning and invention. In Sue T. Parker & Kathleen R. Gibson (Eds.), *"Language" and intelligence in monkeys and apes: Comparative developmental perspectives* (pp. 540–578). New York: Cambridge University Press.

Halliday, Michael A.K. & Ruqaiya Hasen (1976). *Cohesion in English*. London: Longman.

Herman, Louis M., Stan A. Kuczaj, & Mark D. Holder (1993). Responses to anomalous gestural sequences by a language-trained dolphin: Evidence for processing of semantic relations and syntactic information. *Journal of Experimental Psychology: General, 122*, 184–194.

Jensvold, Mary Lee A. & R. Allen Gardner (2000). Interactive use of sign language by cross-fostered chimpanzees. *Journal of Comparative Psychology, 124*, 335–346.

Keenan, Elinor O. (1977). Making it last: Repetition in children's discourse. In Susan Ervin-Tripp & Claudia Mitchell-Kernan (Eds.), *Child discourse* (pp. 125–138). New York: Academic Press.

Kegl, Judy A. & Ronnie B. Wilbur (1976). When does structure stop and style begin? Syntax, morphology, and phonology vs. stylistic modulation in American Sign Language. In Salikoko Mufwene, Carol Walker, & Stanford Steever (Eds.), *Papers from the Twelfth Regional Meeting, Chicago Linguistic Society.* Chicago: University of Chicago.

Klima, Edward S. & Ursula Bellugi (1979). *The signs of language*. Cambridge, MA: Harvard University Press.

Leonard, Laurence B. (1976). *Meaning in child language.* New York: Grune & Stratton.

Lewontin, Richard (2001). *The triple helix: Gene, organism, and environment.* New York: Cambridge University Press.

Livingston, Sue (1983). Levels of development in the language of deaf children: ASL grammatical processes, Signed English structures, semantic features. *Sign Language Studies, 40*, 193–285.

Matsuzawa, Tetsuro (1985). Color naming and classification in a chimpanzee. *Journal of Human Evolution, 14,* 283–291.

Moerk, Ernest L. (1983). *The mother of Eve — As a first language teacher.* Norwood, NJ: Ablex Publishing Corp.

Mohay, Heather (1982). A preliminary description of the communication systems evolved by two deaf children in the absence of a sign language model. *Sign Language Studies, 34*, 73–90.

Moore, Timothy E. (Ed.) (1973). *Cognitive development and the acquisition of language.* New York: Academic Press.

Nelson, Karen (1973). Structure and strategy in learning to talk. *Monograph of the Society for Research in Child Development, 38* (1/2, Serial No. 149), 1–137.

Pepperberg, Irene M. (1992). Proficient performance of a conjunctive, recursive task by an African Gray Parrot. *Journal of Comparative Psychology, 106,* 295–305.

Pinker, Steven (1994). *The language instinct.* New York: W. Morrow and Co.

Premack, David (1971). Language in chimpanzee? *Science, 172,* 808–822.

Reich, Peter A. (1986). *Language development.* Englewood Cliffs, NJ: Prentice Hall.

Rimpau, James B., R. Allen Gardner, & Beatrix T. Gardner (1989). Expression of person, place, and object in ASL utterances of children and chimpanzees. In R. Allen Gardner, Beatrix T. Gardner, & Thomas E. van Cantfort (Eds.), *Teaching sign language to chimpanzees* (pp. 240–268). Albany, NY: SUNY Press.

Robert, Jason S. (2004). *Embryology, epigenesis and evolution.* Cambridge, MA: Harvard University Press.

Rondal, J. A. & D. Defays (1978). Reliability of mean length of utterance as a function of sample size in early language development. *The Journal of Genetic Psychology, 133*, 305–306.

Rosier, Elyse A. (1994). The functions of repetition in American Sign Language narratives and conversation. Unpublished master's thesis, University of Colorado.

Rumbaugh, Duane M. (Ed.) (1977). *Language learning by a chimpanzee.* New York: Academic Press.

Savage-Rumbaugh, E. Sue (1984). Verbal behavior at a procedural level in the chimpanzee. *Journal of the Experimental Analysis of Behavior, 41,* 223–250.

Shatz, Marilyn & Rochel Gelman (1973). The development of communication skills: Modificants in the speech of young children as a function of listener. *Monographs of the Society for Research in Child Development, 38 (5),* 1–38.

Shepherd, Susan C. (1990). Functions of repetition: Modulation in narrative and conversational discourse. In Jerold A. Edmondson, Crawford Feagin, & Peter Muhlhausler (Eds.), *Development and diversity: Linguistic modulation across time and space* (pp. 629–638). Arlington, TX: Summer Institute of Linguistics and University of Texas at Arlington.

Shepherd, Susan C. (1994). Grammaticalization and discourse functions of repetition. In Barbara Johnstone (Ed.), *Repetition in discourse interdisciplinary perspectives* (Vol. 1, pp. 221–229). Norwood, NJ: Ablex Publishing Corp.

Shaw, Heidi L. (2000). Gaze direction in conversational interactions of chimpanzees. Unpublished doctoral dissertation, University of Nevada, Reno.

Siegel, Gerald M. (1962). Interexaminer reliability for mean length of reply. *Journal of Speech and Hearing Research, 5,* 91–95.

Snow, Catherine E. (1972). Mothers' speech to children learning language. *Child Development, 43,* 549–565.

Snow, Catherine E. & Charles A. Ferguson (Eds.) (1977). *Talking to children.* Cambridge: Cambridge University Press.

Stamps, Judy (2003). Behavioural processes affecting development: Tinbergen's fourth question comes of age. *Animal Behaviour, 66,* 1–13.

Stokoe, William C. (1960). Sign language structure: An outline of the visual communication systems of the American deaf. *Studies in linguistics (Occasional Papers).* Buffalo, NY: University of Buffalo.

Terrace, Herbert S., Laura Pettito, Richard J. Sanders, & Thomas G. Bever (1979). Can an ape create a sentence? *Science, 206,* 891–902.

Wells, Gordon (1974). Learning to code experience through language. *Journal of Child Language, 1,* 243–269.

Wilcox, M. Jeanne & Elizabeth J. Webster (1980). Early discourse behaviors: An analysis of children's responses to listener feedback. *Child Development, 51,* 1220–1225.

Zadeh, Lofti A. & Janusz Kacprzyk (Eds.) (1992). *Fuzzy logic for the management of uncertainty.* New York: Wiley.

Part III

Gestural communication in human primates

Human twelve-month-olds point cooperatively to share interest with and helpfully provide information for a communicative partner

Ulf Liszkowski
Max-Planck-Institute for Evolutionary Anthropology, Leipzig

This paper investigates infant pointing at 12 months. Three recent experimental studies from our lab are reported and contrasted with existing accounts on infant communicative and social-cognitive abilities. The new results show that infant pointing at 12 months already is a communicative act which involves the intentional transmission of information to share interest with, or provide information for other persons. It is argued that infant pointing is an inherently social and cooperative act which is used to share psychological relations between interlocutors and environment, repairs misunderstandings in proto-conversational turn-taking, and helps others by providing information. Infant pointing builds on an understanding of others as persons with attentional states and attitudes. Findings do not support lean accounts on early infant pointing which posit that it is initially non-communicative, does not serve the function of indicating, or is purely self-centered. It is suggested to investigate the emergence of reference and the motivation to jointly engage with others also before pointing has emerged.

Pointing in human primates

Pointing is foundational to human communication and has the primary function of indicating an object or location in space (e.g., Kita, 2003; Brinck, 2004). However, pointing would not be foundational to human communication if its indicative function was not understood as being *for someone*. Pointing is not an individualistic goal-directed action upon the physical environment, like, for example, reaching for or manipulating an object. Instead, human pointing is a cooperative activity between individuals, a communicative act, which involves a sender's communicative intention to both transmit information and have a person receive the information on the basis of the sender's communicative intention (Sperber & Wilson, 1995). Bratman (1992) has convincingly argued that human cooperation involves helping the other continue his part in a joint activity. Pointing as cooperative act can also be helpful. For example, in the course of joint cooperative activities we frequently point fast and effortlessly to provide information for a person, to help her overcome misunderstandings about a referent, or to help her find items she might be looking for.

Without the given context of a point it would be impossible to derive a meaning beyond its indication. For a point to communicate meaning it has to be embedded in a context which is construed by the interlocutors' relations towards each other and the environment. Therefore, interlocutors must be able to understand the relations between each other and the environment, i.e. to share each other's attention (Tomasello, 1999), or to mutually manifest knowledge (Sperber & Wilson, 1995). Pointing thus provides a means for a 'meeting of minds' in the external environment. Social-cognitively, human communicative pointing in a shared context requires an understanding of the indication as being about a referent, and an understanding of the interlocutors' psychological relations towards each other and the referent.

Interestingly, non-human primates in captivity also produce the pointing gesture (Leavens & Hopkins, 1998) although it is claimed that they lack the social-cognitive abilities necessary for this (Povinelli, Bering, & Giambrone, 2003), have problems in understanding the communicative intent of pointing (Itakura, Agnetta, Hare, & Tomasello, 1999), and clearly do not engage in what might resemble human-like communication (Tomasello, 2006). It might seem then that human 1-year-olds too point without a deeper social-cognitive understanding, because it has been claimed to emerge only later around 3 to 4 years (see Wellman, Cross, & Watson, 2001). But human 1-year-olds have no problems understanding the communicative intent of pointing (Behne, Carpenter, & Tomasello, 2005; Camaioni, Perucchini, Bellagamba, & Colonnesi, 2004) and they become competent linguistic communicators fairly early, both of which already reveals some kind of mental understanding. And human pointing has been related to symbolic communication theoretically (Werner & Kaplan, 1963) and to language acquisition empirically (e.g., Goldin-Meadow & Butcher, 2003). Therefore, it is questionable whether pointing in human ontogeny really resembles ape pointing and initially does not reflect any mental understanding, or whether it already bears cognitive and motivational properties of uniquely human communicative pointing when it has just emerged.

In resolving this question, this paper reports three recent experimental studies from our lab (Liszkowski, Carpenter, Henning, Striano, & Tomasello, 2004;

Liszkowski, Carpenter, & Tomasello, 2007; Liszkowski, Carpenter, Striano, & Tomasello, 2006) which investigated in detail communicative motives and social-cognitive abilities of pointing in 12-month-old human infants. New findings show that infant pointing already is a communicative act, even before language has emerged. It is motivated by cooperatively sharing interest with or helpfully providing information for other persons and builds on infants' understanding of others as persons with attentional states and attitudes. These new findings are contrasted with existing accounts of infant pointing and underlying communicative and social-cognitive abilities which are reviewed below.

Review of infant pointing

Bates, Camaioni and Volterra (1975) first described infant pointing in a longitudinal investigation of infant communication. Following Austin's (1962) speech act theory, they proposed a developmental sequence in infant communication from perlocutionary to illocutionary to locutionary acts. Pointing was claimed to correspond to the illocutionary stage, revealing the intent to signal to a recipient. However, Bates et al. (1975) also reported pointing which they classified as non-communicative, based on the absence of gaze alternation to a recipient. They interpreted non-communicative pointing as a precursor to communicative pointing.

Since then, pointing has been suggested to become intentionally communicative only later in development, after its initial emergence, at around 15 months (Desrochers, Morissette, & Ricard, 1995), possibly through caregivers' communicative responses. As a criterion for intentional communication, research has usually relied on gaze alternation to the recipient. Methodologically, however, it might be misleading to use looks as the only main criterion to assess communicative intent. For example, infants might alternate gaze simply to check on the other person, without communicative intent. And absence of gaze alternation would not necessarily mean an absence of communicative intent — because infants might simply assume that adults understand the behavior as communicative, or rely on auditory instead of visual information. Other criteria for intentional communication are whether it is done for somebody and whether persistence and flexibility in signal-use occur when the recipient does not react accordingly (see also Tomasello & Call, 1997). The three new studies from our lab will show that 12-month-olds' pointing already is intended to be communicative.

But infants also point for themselves. Delgado, Gómez, and Sarria (1999) observed that infants point even when they are alone in a room, without audience. However, Delgado and colleagues (2002, 2004) also showed that preschoolers at

3 and 5 years still point for themselves. Therefore, it is unlikely that such pointing for self is a precursor to communicative pointing — because it does not disappear when children already point communicatively. Instead, such pointing might serve a function similar to that of private speech (Vygotsky, 1978)[1]. This interpretation is supported by DeLoache, Cassidy, and Brown (1985) who found that infants sometimes use pointing as a mnemonic strategy. Further, Bruner (1983) described an infant pointing for self without perceivable referent as "locating in his 'present' space an object recalled from memory" (p. 76). Pointing for self, then, seems to coexist with rather than develop into communicative pointing in infancy. It might even be hypothesized that pointing for self develops only after the communicative function of pointing already is established.

Communicative pointing in infancy has been claimed to involve rather self-centered motives, like using the adult as a tool to obtain an object ('proto-imperative'), or the object as a tool to obtain adult attention ('proto-declarative'), with an understanding of causality corresponding to the Piagetian level 5 of sensori-motor development (Bates et al., 1975). Subsequently, these two types of communicative pointing have received different interpretations in terms of their communicative and cognitive complexity (see Brinck, 2004).

Imperative pointing has typically been interpreted on a leaner, more behavioristic account, and declarative pointing on a richer, more mentalistic account (Camaioni, 1993). For example, Vygotsky (1978) claimed pointing to be a ritualized behavior through adults' repeated interventions to failed attempts of reaching, and Wundt (cited in Werner & Kaplan, 1963) described it as an "abbreviated grasp". But Franco and Butterworth (1996) found reaching and pointing to serve different functions in development, and Masataka (2003) showed reaching and pointing to be developmentally not associated. Nevertheless, presumably because imperative pointing is more about spurring others into action, it has been interpreted as a self-centered instrumental act, at most revealing some causal understanding of others' agency ('agent of action', Brinck, 2004; Camaioni, 1993; Gómez, Sarria, & Tamarit, 1993). The case that apes in captivity and children with autism can point imperatively despite the lack of a necessary understanding of others' mental agency (Tomasello, 2006; Baron-Cohen, 1989) has lent support to this interpretation. However, adults point imperatively with an understanding of others' mental agency and, without other evidence, it is at least possible that typically developing infants point in this way as well; we simply do not know.

Declarative pointing, in contrast, has been taken to reflect sensitivity to others' mental agency. It is less about spurring someone into action than changing a person's attentional state (e.g., Baron-Cohen, 1991; Bretherton, McNew, & Beeghly-Smith, 1981; Tomasello, 1995). It has been claimed to be motivated by sharing attention, a motivation manifest also in other Joint Attention behaviors such as gaze following, social referencing, giving, showing and imitating (Tomasello, 1999), all of which emerge as a cluster around infants' first birthday and are related to the onset of language (Carpenter, Nagell, & Tomasello, 1998). In addition, apes (however they are raised) and children with autism do not point declaratively, presumably because they lack the necessary cognitive ability or are unmotivated to do so.

Contrary to rich accounts of early declarative pointing, some researchers have expressed skepticism that declarative pointing, when it has just emerged, involves an understanding of others' mental agency (Carpendale & Lewis, 2004; Gómez et al., 1993; Moore & Corkum, 1994). For example, Gómez et al. (1993) have suggested that infants simply understand a recipient's behavioral relation to a referent when they point. More recently, Moore and D'Entremont (2001) have claimed that 12-month-olds do not point to direct others' attention. In an experiment, they found that 12-month-olds pointed equally often at an event, irrespective of whether an adult already was looking at it. They concluded that infants initially point only to obtain attention to the self.

Camaioni, in her work on infant pointing (1975–2004) took an intermediate position putting forward both lean and rich accounts. Like Brinck (2004), she separated imperative from declarative pointing, and suggested a developmental décalage between these two types. On her account, imperative pointing emerged before declarative pointing which she interpreted as a social-cognitive transition from an understanding of other persons as 'agents of action' to 'agents of contemplation' (Camaioni, 1993). Whereas she suggested that early declarative pointing revealed an understanding of others' intentionality, she claimed that early imperative pointing did not require such an understanding. In her latest work (Camaioni et al., 2004) she empirically addressed this hypothesis, showing that imperative pointing was more frequent than declarative pointing among infants who had just begun pointing, and that declarative, but not imperative pointing was developmentally associated with passing Meltzoff's (1995) task of imitating failed attempts.

There are thus rich and lean accounts of infants' early pointing. Minimally, there is agreement that later in development, around the end of infants' second year, declarative pointing is about directing others' attention (Moore & D'Entremont, 2001) and about informational exchange (Franco & Gagliano, 2001). However, infants begin pointing a year earlier, around 12 months (Leung & Rheingold, 1981). It is thus not clear what infants do when they have just begun pointing. Existing accounts of 12-month-olds' pointing have not systematically been tested and consequently we lack necessary evidence on why young infants point. In overcoming this gap, this paper reports three of our recent studies which addressed the motives and social-cognitive abilities underlying infant pointing at 12 months.

Twelve-month-olds point to share attention and interest

In a recent study, Liszkowski et al. (2004) tested why 12-month-olds point in a classical declarative context (see Figure 1). Infants were presented with 10 interesting events, like hand puppets appearing behind a large screen at a distance or lights flashing, and a female experimenter (E) reacted consistently in one of four specific ways to each infant's pointing. We were interested whether infants would be more satisfied with one over another reaction and whether they would modify their behavior as a function of E's reaction to their pointing. For example, we measured how often infants would point in the different social contexts, whether they would 'repair' their message and repeat pointing to the same referent more if E did not react in the expected way, and whether their looking behavior to E would differ across situations. Specifically, we tested four hypotheses on what infants might want when they point declaratively. In a *Joint Attention* condition, E responded to an infant's points by alternating gaze between the event and the infant and emoted positively about it on the hypothesis that infants want to share attention and interest. In the *Face* condition, E never looked at the event and instead attended to the infant's face and emoted positively to it, on Moore and D'Entremont's (2001) hypothesis that infants do not want to direct attention but just want to obtain attention to the self. In the *Event* condition, E only attended to the events, on the

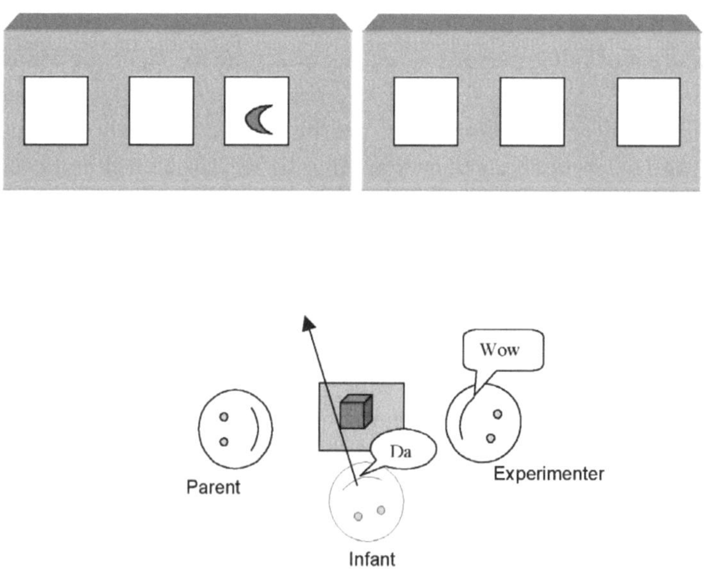

Figure 1. Study 1. Schematic drawing of the set-up. Back: screen with window-openings and protruded stimulus; front: infant in high-chair with an attached table and small, attached toy.

Table 1. Study 1. Summary of Main Results. '+' Indicate Statistically Higher Numbers than '–'. Means in Parentheses.

	Prop. of trials with point	# of points per trial	# of looks to E per trial
Joint Attention	+ (0.7)	– (1.07)	– (0.28)
Face	– (0.5)	+ (1.23)	– (0.33)
Event	– (0.5)	+ (1.23)	+ (0.77)
Ignore	– (0.4)	+ (1.19)	– (0.44)

hypothesis that infants just want to direct attention and nothing else, and in the *Ignore* condition E attended neither to the infant nor to the event, on the hypothesis that infants might point non-communicatively, for themselves.

Table 1 summarizes the main statistically significant differences between conditions. The overall finding was that infants point to share attention and interest. First, infants were more satisfied in the *Joint Attention* condition and pointed on significantly more trials in that condition compared to the other three. Second, infants were not satisfied in the *Face* condition, when E only emoted positively to them. In that condition, although E emoted as positively to the infant as in the *Joint Attention* condition, infants repeated their pointing to the same referent significantly more often than when she shared attention to it. In the *Face* condition, infants thus attempted to redirect E's attention to the event. Third, infants were not satisfied in the *Event* condition either. When E only attended to the event and did not comment back, infants also repeated their pointing more within a trial than in the *Joint Attention* condition. In addition, they looked more to E than in any other condition, presumably because they expected E to comment back.

Results show, first, that infants point intentionally communicatively and tailor their communicative behavior to different social responses. Second, in a declarative context, infants point to share their attention and interest with a communicative partner. Sharing attention and interest involves both (i) directing the other person's attention *and* (ii) receiving a comment about the mutually attended to event; neither alone is sufficient.

We have recently followed up these results in a new experiment (Liszkowski et al., 2007) and investigated the two components, directing attention and receiving a comment, in more detail. Again we used a declarative context (see Figure 2) to elicit pointing, and a male experimenter (E) sitting in front of the infant with the back turned to the stimuli responded in one of four specific ways. Specifically, we were interested which of E's reactions might satisfy the infant's motive to share attention and interest. Therefore we systematically violated infants' expectations of E's attention and of his comment. E either did not share the infant's attention, i.e. he did not refer to what the infant pointed at, or he did not share the infant's

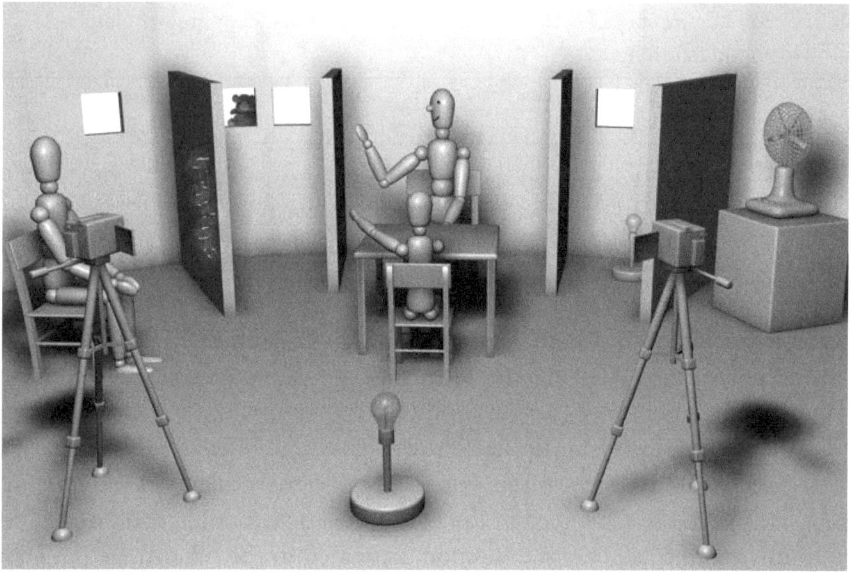

Figure 2. Study 2. Schematic drawing of the set-up with barriers.

interest, i.e. he commented uninterestedly about the referent. First, we wanted to know whether infants would be satisfied when E simply oriented behaviorally into the direction of the referent without actually attending to it (a barrier obstructed his line of sight). Second, we wanted to know whether the adult needed to comment positively, or whether a neutral comment would suffice. We thus controlled two components of E's reaction to infants' pointing: (i) the referent of E's attention and (ii) E's attitude toward the referent, as expressed in his comment. This resulted in four conditions. In the *Joint Attention* condition, E attended to the infant's referent and emoted positively about it (but never named it), emphasizing his attention toward it by turning head and body towards it and slightly extending his arm, palm up into its direction. In the *Misunderstanding* condition, E reacted in the same way, except that a barrier obstructed his line of sight to the infant's referent and E mistakenly referred to an insignificant piece of paper attached to the barrier. In the *Uninterested* condition, there was no barrier and E reacted as in *Joint Attention*, except that he commented neutrally about the referent, stating his disinterest in it. The *No Sharing* condition involved the same barriers as in the *Misunderstanding* condition and E commented neutrally as in the *Uninterested* condition to an alternative referent on the barrier.

Table 2 summarizes the main results. First, as in the previous study, infants were more satisfied in the *Joint Attention* condition, pointing on more trials in that condition than in the other three. Second, in the *Misunderstanding* condition, when E emoted as positively as in the *Joint Attention* condition and behaviorally

Table 2. Study 2. Summary of Main Results. '+' Indicate Statistically Higher Numbers than '–'. Means in Parentheses.

	Prop. of trials with point	Prop. of trials with repetitions	# of looks to E per trial
Joint Attention	+ (0.9)	– (0.3)	– (0.7)
Misunderstanding	– (0.7)	+ (0.5)	+ (1.7)
Uninterested	– (0.6)	– (0.2)	+ (1.9)
No Sharing	– (0.6)	– (0.3)	+ (2.2)

Table 3. Study 2. Qualitative Differences of Point Repetitions between *Joint Attention* and *Misunderstanding*. '+' Indicate Statistically Higher Numbers than '–'. Means in Parentheses.

	Point repetitions		
	Latency in sec. to 2nd point	# of looks to E between 1st and 2nd point	# of vocalizing during 2nd point
Joint Attention	– (4.9)	– (0.49)	– (0.57)
Misunderstanding	+ (6.6)	+ (1.27)	+ (0.82)

oriented into the same direction but referred to the barrier instead of the referent, infants were not satisfied. In that condition (*Misunderstanding*) they persisted in their message and repeated pointing to the referent within trials more than when E attended to the referent (*Joint Attention*). In addition, these point repetitions were accompanied by significantly more gaze alternation to E and vocalizations, and less impulsive than in the *Joint Attention* condition (see Table 3). Third, in the *Uninterested* condition, when E attended to the referent just as in the *Joint Attention* condition but commented neutrally about it, infants did not repeat their pointing within trials. Although infants were overall less satisfied with a neutral comment response — they pointed on fewer trials than in the *Joint Attention* condition — they did not repeat pointing within trials in order to receive a different response.

Results show, first, that infants point to direct another person's attention to the event which they point at. This is in line with the results of the *Face* condition in the first study. Importantly, results show that infants were not satisfied when the recipient oriented only behaviorally and simply turned into the direction of the referent, even when he emoted positively. This means that infants do not point simply to direct a recipient's external bodily behavior, or simply to elicit a positive comment. Instead, infants point to direct the other person's attentional state to what they themselves attend to. Second, results show that in this context, infants prefer a positive over a neutral comment about a mutually attended to event. However, in contrast to the *Event* condition of the first experiment, in which infants repeated pointing when they did not receive any comment, in the *Uninterested* condition, when the comment was not the preferred one, infants did not repeat their pointing to receive a different reaction from E. This shows that infants do not point simply to request a positively emoted comment about an event mutually attended to, like they request, for example, when imperatively pointing to obtain a cookie. Instead, infants' pointing in such a context rather resembles an offer (see Bruner, 1983), to mutually engage about an event and share interest in it with an interested partner.

Taken together, findings of these two recent studies strongly support the interpretation that infants point in a context in which interesting things happen or appear to share their attention and interest with a communicative partner. Further, findings do not support alternative, leaner hypotheses of infant pointing in such a context. First, contrary to Desrochers et al. (1995), our findings clearly show that 12-month-olds already point intentionally communicatively and not simply for themselves. Second, in contrast to Moore and D'Entremont's (2001) hypothesis, infants do not point simply to obtain an adult's attention to themselves. They are more satisfied when the adult also attends to the referent which they point at (in fact, then the adult attends even less to the infant because he divides his attention between infant and object). Third, it is unlikely that infants want the adult only to behaviorally orient and relate to a referent. Instead, they really want him to see and attend to what they attend to. And finally, infants do not simply request an adult to positively emote when mutually attending to an event, as if they imperatively requested a positive object-related comment. Instead, infants point to initiate a joint bout by offering to share their interest in an event.

Theoretically, it is possible that infants also point interrogatively, to receive information about the referent, e.g. its valence or its word label. On such an account one would expect infants to point irrespective of how E commented (i.e., positively or neutrally), as long as he provided information. However, in the current situation infants selectively preferred a positive comment over a neutral. Further, in the second study, they were never provided with the actual word label of the referent, and this did not affect their pointing substantially compared to the first study. Whether and when infants point interrogatively thus remains an empirical question — in the current context they pointed to share attention and interest.

Social-cognitively, first, findings show that infant pointing reveals an understanding of other persons' attention. Results show that infants differentiate between conditions in which the recipient is and is not attending to what they point at. This is consistent with recent results on gaze and point-following which show that 12-month-olds understand what other persons are looking at (Deák, Flom, & Pick, 2000) even in the presence of distractors and when the target is behind their own visual field. Infants follow others' looks and points over their first year of life, attending to what others are attending to (e.g., D'Entremont, Hains, & Muir, 1997) and come to reverse roles when they point at objects to direct others' attention to what they attend to. Twelve-month-olds thus understand that other persons' attention can be aligned to an object, just as their own can be.

Social-cognitively, second, infants' motive to share interest suggests that they conceive of others as persons with attitudes towards the environment. Just as infants come to understand that they can follow others' attention, infants also experience over their first year of live that other persons express to them their psychological relations, that is, their attitudes towards the environment (see also Hobson, 1994). For example, when adults direct infants' attention they also express their attitude about the mutually attended to referent towards the infant, which is often one of positive interest (in other situations adults might also direct attention to provide information, e.g. to tell them where a toy is [see Behne et al., 2005], or to request an object when it is in the infant's possession [see Camaioni et al., 2004]). Studies on social referencing have shown that infants can link a person's comment selectively to objects (Moses, Baldwin, Rosicky, & Tidball, 2001). However, social referencing is more about the referent, to discern ambiguity of situations or objects, than it is about sender and recipient. Therefore, it has been suggested that infants apply the adult's comment only to the object, as its valence, and do not understand the comment as the adult's psychological relation to the object (Egyed, Kiraly, & Gergely, 2004). Pointing to share attention and interest, however, is not only about the referent. It is a joint, cooperative activity, for both sender *and* recipient. Results show that the way an adult comments to the infant about a referent influences the infant's incidence of subsequent offers to share attention and interest. When pointing, infants conceive of a recipient's comment as expressing his attitude about the referent. If the attitude is similar to their own, they share this attitude which, in the case of declarative pointing, is one of positive interest.

Motivationally then, pointing at 12 months already is an inherently social communicative act, intended for both sender and recipient, and not a solitary, self-centered activity of the infant. Infants' 'repair' of misunderstandings resembles conservational turn-taking structures and is helpful for a recipient in understanding the communicative intent. The cooperative structure underlying early communicative acts may be interpreted as uniquely human (Tomasello et al., 2005) and as being at the roots of human sociality (see Enfield & Levinson, 2006).

Twelve-month-olds point informatively to help others

In another recent study (Liszkowski et al., 2006), we further explored infants' motives for directing other persons' attention when pointing. Infants' repair of misunderstandings can be interpreted as helpful in proto-conversational turn-taking. As Bratman (1992) has argued, human shared cooperative activity involves helping a partner in keeping up his part. We therefore investigated whether infants might indeed, in some situations, be motivated to help a recipient when pointing. As adults we often point to help others by providing information, maybe even more than we point to share interesting events. For example, when we see a person searching for something (e.g., walking or turning around, looking into various directions, inspecting various locations — looking *for something*), we often help that person by pointing out for her what she is looking for (or what we believe she might be looking for). Clearly, in such a situation we neither point to request the referent for ourselves, nor to express our excitement about it. Instead, one points to inform a person about an object's location, to help her find it. Interestingly, we helpfully point things out for persons who we do not know and might never see again (e.g., on the street, in a concert, on a train, etc.), without direct benefit to ourselves. Such helping behavior can be interpreted as part of a cooperative activity (being *together* on a street, in a concert, in the same political party, in a social psychology experiment; see Clark, 2006). It might also be interpreted as altruistically helping a stranger, although there are arguably no 'high' costs involved (e.g., risk of life), and it might not extent to 'non-peer' strangers (e.g., radical opponents of one's favorite soccer team, political party, etc.). In any case, such type of pointing is informative in helping another person find what she is looking for. The motive of informative pointing thus departs from the classical dichotomy of imperative and declarative pointing in infancy because it is neither *to obtain an object for self* nor *to share interest* in it. Instead, the motive is to *helpfully provide information for the other.*

In this study we investigated whether infants point informatively. In the main experiment, a female experimenter (E) demonstrated on each of twelve trials an action to the infant, which always involved one of two objects (the target). Then, both objects (target and distractor) disappeared out of E's but not the infant's view, for example they were dropped accidentally, displaced on a shelf behind E, or used up while replica objects remained visible to the infant. After the disappearance, E attempted to repeat the action and began searching for the target object. She first looked for herself, then emphasized her search with an unspecific verbal cue ("where is it?") and then explicitly asked the infant using the object label ("[Name], where is the [target]?").

Results were that infants pointed in such a situation, even when potentially interesting sounds or movements of the referent were absent, or when there was no displacement at all (e.g., when the objects were used up). Infants pointed significantly more to the target which E was looking for than to a distractor simultaneously displaced (see Figure 3). Requestive accompaniments or repeated pointing after E had retrieved the object were rare and infants pointed mostly before E verbally asked about the object.

Findings show that infants point informatively to provide information for another person. Pointing in such a context is not so much about a sender's relation to a stimulus, to share it, than about a recipient's relation to it, to help him. The results show that infants do not only direct an adult's attention in response to externally salient, exciting events (e.g. Butterworth et al., 2002) but also because of a recipient's relation to the referent, e.g. to find it. There are good reasons to believe that infants did not simply request the object for themselves, nor to have the adult simply do an action. First, requestive accompaniments typically associated with imperative pointing were very rare, and when E had retrieved the object, infants rarely attempted to obtain it. And second, most of the actions were not particularly interesting, without specific effects, and infants were never involved in them (they

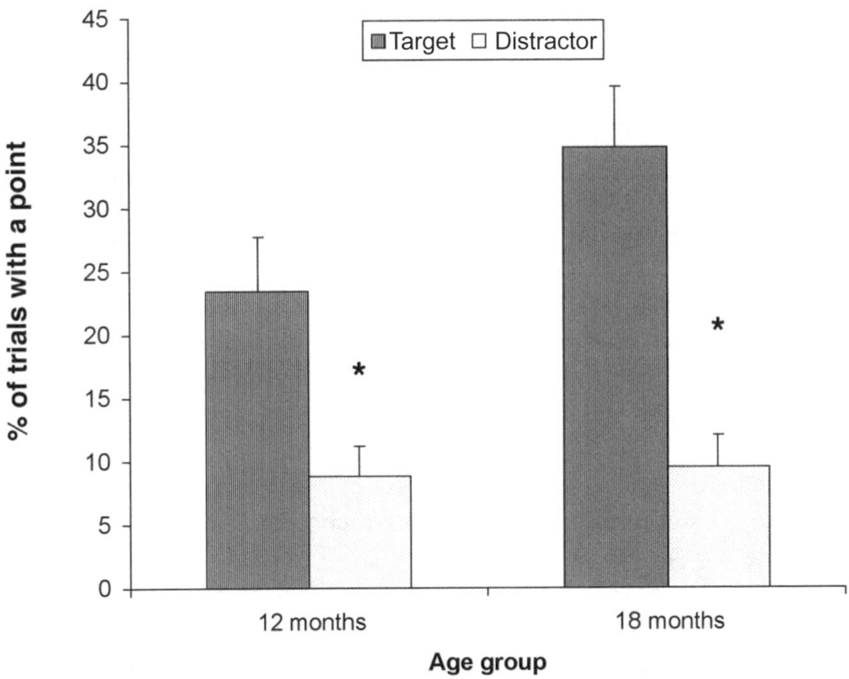

Figure 3. Study 3. Percent of trials with a point to target and distractor. * indicate statistically significant differences ($p < .05$).

simply watched E). In addition, in a prior experiment (reported in Liszkowski et al., 2006), the objects which disappeared were previously not involved in a particular goal-directed action and infants still pointed.

Social-cognitively, first, informative pointing reveals infants' understanding of a recipient's relation to an object. To point informatively, one must understand what the addressee wants. In this context, infants understood E's intention to find an object and continue an action, and helped to complete his goal by pointing out the object. Results thus reveal an understanding of others' goals and intentions to be present in infant pointing at twelve months. Second, in line with the two previous studies on declarative pointing, our findings of informative pointing foster the interpretation that infants conceive of others as persons with attentional states who can sometimes lack relevant information. This is consistent with recent results by Tomasello and Haberl (2003) who showed that 12-month-olds can discern what is new for someone else. The findings extend previous work on older children's informational exchange in declarative (Franco & Gagliano, 2001) and imperative (O'Neill, 1996) contexts, and suggest that infants' informative pointing already reflects an understanding of persons having information states about the environment.

Motivationally, findings are that infants provide information freely without concern for immediate personal benefit which reveals the prosocial motivation of helping a communicative partner. This extends previous work by Rheingold (1982) who showed that older infants participate in joint actions, like cleaning up or opening things. In this study, physical assistance was not required and it may be that humans are especially inclined to help communicatively. Helping others is an integral part of human cooperation (Bratman, 1992). This study therefore supports the interpretation that infant pointing is a humanly cooperative, communicative act. The motivation to provide information and help other persons might be seen as the initial ontogenetic emergence of the uniquely human ability to teach and instruct other persons. Parts of such cooperative instructing are suggested to emerge early in ontogeny, before language, in communicative behaviors like informative pointing.

Conclusion: Infant pointing is a cooperative, communicative act

New findings of three recent experimental studies presented here (Liszkowski et al., 2004, 2006, 2007) show that human pointing, when it has just emerged, is a communicative act which involves the intentional transmission of information by directing another person's attention to an indicated object or event. Liszkowski et al.'s new approach to pointing which considers motives instead of a general differentiation between different types of pointing shows that infants' motives of pointing are humanly cooperative in nature. Specifically, infants' pointing at 12 months is motivated by mutually sharing interest in an event with a communicative partner. Moreover, in addition, infants point to help by providing information for another person, a motive which has previously not been investigated even though it is very common in adult everyday pointing. Findings thus do not support lean accounts of early infant pointing which have suggested that it is initially non-communicative, does not serve the function of indicating, or is purely self-centered. Therefore, it also seems that human pointing in ontogeny is already fundamentally different in its function and use to the gestures exhibited by apes in captivity. Future research needs to empirically investigate developmental antecedents of infant pointing by investigating the role of early social interaction in the ontogenetic emergence of reference and, interwoven with the emergence of referential behaviors, the motivation to cooperate with and help each other.

Acknowledgements

This paper is in memory of Luigia Camaioni. I am thankful to Emily Wyman and Mike Tomasello for comments on a previous version. Parts of this paper were presented at the Workshop for Gestural Communication in Non-human and Human Primates, Leipzig, Germany, 2004, and at the Wenner-Gren Foundation for Anthropological Research, Symposium 134 "Roots of Human Sociality: Culture, Cognition, and Human Interaction", October 2–9, 2004, Duck, North Carolina, US.

Note

1. I am thankful to B. Delgado and J. C. Gómez for insightful discussions on this point.

References

Austin, John L. (1962). *How to do things with words*. New York: Oxford University Press.

Baron-Cohen, Simon (1989). Perceptual role-taking and protodeclarative pointing in autism. *British Journal of Developmental Psychology, 7* (2), 113–127.

Baron-Cohen, Simon (1991). Precursors to a theory of mind: Understanding attention in others. In Andrew Whiten (Ed.), *Natural theories of mind: Evolution, development and simulation of everyday mindreading* (pp. 233–251). Oxford (UK): Blackwell.

Bates, Elizabeth, Luigia Camaioni, & Virginia Volterra (1975). The acquisition of performatives prior to speech. *Merrill-Palmer Quarterly, 21*(3), 205–226.

Behne, Tanya, Malinda Carpenter, & Michael Tomasello (2005). One-year-olds comprehend the communicative intentions behind gestures in a hiding game. *Developmental Science, 8*, 492–499.

Bratman, Michael (1992). Shared cooperative activity. *The Philosophical Review*, 101, 327–341.

Bretherton, Inge, Sandra McNew, & Marjorie Beeghly-Smith (1981). Early person knowledge as expressed in gestural and verbal communication: When do infants acquire a "theory of mind"? In Michael E. Lamb & L. R. Sherrod (Eds.), *Infant social cognition: Empirical and theoretical considerations*. Hillsdale, NJ: Erlbaum.

Brinck, Ingar (2004). The pragmatics of imperative and declarative pointing, *Cognitive Science Quarterly*, 3 (4).

Bruner, Jeromy (1983). *Child's talk*. New York; London: Norton & Company.

Butterworth, George, Fabia Franco, B. McKenzie, L. Graupner & B. Todd (2002). Dynamic aspects of visual event perception and the production of pointing by human infants. *British Journal of Developmental Psychology, 20*, 1–24.

Camaioni, Luigia (1993). The development of intentional communication: A re-analysis. In Jacqueline Nadel & Luigia Camaioni (Eds.), *New perspectives in early communicative development* (pp. 82–96). London: Routledge.

Camaioni Luigia, Perucchini Paola, Bellagamba Francesca, & Cristina Colonnesi (2004). The role of declarative pointing in developing a theory of mind. *Infancy, 5*, 291–308.

Carpendale, Jeremy & Charlie Lewis (2004). Constructing an understanding of mind: The development of children's social understanding within social interaction. *Behavioral and Brain Science, 27*(1), 79–96.

Carpenter, Malinda, Katherine Nagell, & Michael Tomasello (1998). Social cognition, joint attention, and communicative competence from 9 to 15 months of age. *Monographs of the Society of Research in Child Development, 63*(4), Serial No. 176.

Clark, Herbert H. (2006). Social actions, social commitments. In Nick Enfield & Steve Levinson (Eds.), *The roots of human sociality: Culture, cognition, and interaction*. Oxford: Berg.

Deák, Gedeon O., Ross A. Flom, & Anne D. Pick (2000). Effects of gesture and target on 12- and 18-month-olds' joint visual attention to objects in front of or behind them. *Developmental Psychology, 36*(4), 511–523.

Delgado, Begoña, Juan C. Gómez, & Encarnacion Sarriá (2004). Is pointing more than a communicative gesture? A study about the role of pointing in regulating one's own attention. *Poster presented at the 18th Biennal meeting of the International Society for the Study of Behavioural Development, Gent, Belgium.*

Delgado, Begoña, Juan C. Gómez, & Encarnacion Sarriá (2002). Can young children use their pointing gestures as a private tool for regulating their thought processes? *Poster presented at the 32nd Annual Meeting of the Jean Piaget Society, Philadelphia, USA.*

Delgado, Begoña, Juan C. Gómez, Encarnacion Sarriá (1999). Non-communicative pointing in preverbal children. *Poster presented at the IXth European Conference on Developmental Psychology, Spetses, Greece.*

DeLoache, Judy S., Deborah J. Cassidy, & Ann L. Brown (1985). Precursors of mnemonic strategies in very young children's memory. *Child Development, 56*(1), 125–137.

D'Entremont, Barbara, S.M. Hains, & Darwin Muir (1997). A demonstration of gaze following in 3- to 6-month-olds. *Infant Behavior and Development, 20*(4), 569–572.

Desrochers, Stephan, Paul Morissette, & Marcelle Ricard (1995). Two perspectives on pointing in infancy. In Chris Moore & Philip J. Dunham (Eds.), *Joint attention: Its origins and role in development* (pp. 85–101). Hillsdale NJ: Lawrence Erlbaum.

Egyed, K., I. Kiraly, & G. Gergely (2004). Object-centered versus agent-centered interpretations of attitude expressions. *Poster presented at the International Conference on Infant Studies*, Chicago.

Enfield, Nick & Steven Levinson (Eds.) (2006). *The roots of human sociality: Culture, cognition, and interaction*. Oxford: Berg.

Franco, Fabia & Antonino Gagliano (2001). Toddlers' pointing when joint attention is obstructed. *First Language, 21*(63), 289–321.

Franco, Fabia & George Butterworth (1996). Pointing and social awareness: Declaring and requesting in the second year. *Journal of Child Language, 23*(2), 307–336.

Goldin-Meadow, Susan & Cynthia Butcher (2003). Pointing toward two-word speech in young children. In Sotaro Kita (Ed.), *Pointing: Where language, culture, and cognition meet* (pp. 85–107). Mahwah, NJ, US: Lawrence Erlbaum Associates, Publishers.

Gómez, Juan C., Encarnacion Sarria, & Javier Tamarit (1993). The comparative study of early communication and theories of mind: Ontogeny, phylogeny, and pathology. In Simon Baron-Cohen, Helen Tager-Flusberg et al. (Eds), *Understanding other minds: Perspectives from autism* (pp. 397–426). New York: Oxford University Press.

Hobson, R. Peter (1994). Perceiving attitudes, conceiving minds. In Charlie Lewis & Peter Mitchell (Eds.), *Children's early understanding of mind: Origins and development* (pp. 71–93). Hillsdale, NJ: Lawrence Erlbaum.

Itakura, Shoji, Bryan Agnetta, Brian Hare & Michael Tomasello (1999). Chimpanzees use human and conspecific social cues to locate hidden food. *Developmental Science, 2*, 448–456.

Kita, Sotaro (Ed.) (2003). *Pointing: Where language, culture, and cognition meet*. Mahwah, NJ: Lawrence Erlbaum Associates, Publishers.

Leavens, David A. & William D. Hopkins (1998). Intentional communication by chimpanzees: A cross-sectional study of the use of referential gestures. *Developmental Psychology, 34*(5), 813–822.

Leung, Eleanor H. & Harriet L. Rheingold (1981). Development of pointing as a social gesture. *Developmental Psychology, 17*(2), 215–220.

Liszkowski, Ulf, Malinda Carpenter, Tricia Striano, & Michael Tomasello (2006). Twelve-and 18-month-olds point to provide information. *Journal of Cognition and Development, 7*, 173–187.

Liszkowski, Ulf, Malinda Carpenter, & Michael Tomasello (2007). Twelve-month-old pointing: Reference and attitude in infant pointing. *Journal of Child Language*, 34, 1–20.

Liszkowski, Ulf, Malinda Carpenter, Anne Henning, Tricia Striano, & Michael Tomasello (2004). Twelve-month-olds point to share attention and interest. *Developmental Science, 7*(3), 297–307.

Masataka, Nobuo (2003). From index-finger extension to index-finger pointing: Ontogenesis of pointing in preverbal infants. In Sotaro Kita (Ed.), *Pointing: Where language, culture, and cognition meet* (pp. 69–84). Mahwah, NJ, US: Lawrence Erlbaum Associates, Publishers.

Meltzoff, Andrew N. (1995). Understanding the intentions of others: Re-enactment of intended acts by 18-month-old children. *Developmental Psychology, 31*, 1–16.

Moore, Chris & Barbara D'Entremont (2001). Developmental changes in pointing as a function of attentional focus. *Journal of Cognition & Development, 2*(2), 109–129.

Moore, Chris & Valerie Corkum (1994). Social understanding at the end of the first year of life. *Developmental Review, 14*(4), 349–372.

Moses, Louis J., Dare A. Baldwin, Julie G. Rosicky, & Glynnis Tidball (2001). Evidence for referential understanding in the emotions domain at twelve and eighteen months. *Child Development, 72*(3), 718–735.

O'Neill, Daniela K. (1996). Two-year-old children's sensitivity to a parent's knowledge state when making requests. *Child Development, 67(2).*

Povinelli, Daniel J., Jesse M. Bering, & Steve Giambrone (2003). Chimpanzees' "pointing": Another error of the argument by analogy? In Sotaro Kita (Ed.), *Pointing: Where language, culture, and cognition meet* (pp. 35–68). Mahwah, NJ, US: Lawrence Erlbaum Associates, Publishers.

Rheingold, Harriet L. (1982). Little childrens participation in the work of adults, a nascent prosocial behavior. *Child Development*, 53, 114–125.

Sperber, Dan & Deirdre Wilson (1995). *Relevance: Communication and cognition* (2nd ed.). Oxford: Blackwell.

Tomasello, Michael (2006). Why don't apes point? In Nick Enfield & Steven Levinson (Eds.), *The roots of human sociality: Culture, cognition, and interaction* (pp. 506–524). Oxford: Berg.

Tomasello, Michael (1999). *The cultural origins of human cognition*. Cambridge, MA: Harvard University Press.

Tomasello, Michael (1995). Joint attention as social cognition. In Chris Moore & Philipp Dunham (Eds.), *Joint attention: Its origins and role in development*. Hillsdale: Lawrence Erlbaum Associates.

Tomasello, Michael, Malinda Carpenter, Josep Call, Tanya Behne, & Henrike Moll (2005). Understanding and sharing intentions: The origins of cultural cognition. *Behavioral and Brain Sciences, 28*, 675–691.

Tomasello, Michael & Katharina Haberl (2003). Understanding attention: 12- and 18-month-olds know what's new for other persons. *Developmental Psychology,* 39, 906–912.

Tomasello, Michael & Josep Call (1997). *Primate cognition*. New York, Oxford: Oxford University Press.

Vygotsky, Lev (1978). *Mind in society: The development of higher psychological processes*. Cambridge, MA: Harvard University Press.

Wellman, Henry M., D. Cross, & J. Watson (2001). Meta-analysis of theory-of-mind development: The truth about false belief. *Child Development,* 72(3), 655–84.

Werner, Heinz & Bernard Kaplan (1963). *Symbol formation: An organismic-developmental approach to language and the expression of thought*. New York, Wiley.

From action to language through gesture

A longitudinal perspective

Olga Capirci, Annarita Contaldo, M. Cristina Caselli, and Virginia Volterra

Institute of Cognitive Sciences and Technologies, National Research Council (CNR) — Rome, Italy

The present study reports empirical longitudinal data on the early stages of language development. The main hypothesis is that the output systems of speech and gesture may draw on underlying brain mechanisms common to both language and motor functions. We analyze the spontaneous interaction with their parents of three typically-developing children (2 M, 1 F) videotaped monthly at home between 10 and 23 months of age. Data analyses focused on the production of actions, representational and deictic gestures and words, and gesture-word combinations. Results indicate that there is a continuity between the production of the first action schemes, the first gestures and the first words produced by children. The relationship between gestures and words changes over time. The onset of two-word speech was preceded by the emergence of gesture-word combinations. The results are discussed in order to integrate and support the evolutionary and neurophysiological views of language origins and development.

Introduction

Several studies have emphasized the links between gesture and language in early communicative development of human infants. Some pioneering studies (Bates, Camaioni, & Volterra, 1975; Bates, Benigni, Bretherton, Camaioni, & Volterra, 1979) reported that the onset of intentional communication, between the ages of 9 and 13 months, was marked in part by the emergence of a series of gestures (ritualized request, giving, showing, pointing) that preceded the appearance of first words. These gestures defined as performatives and later deictic, express the child's communicative intent to request or to declare and are used to draw attention to objects, locations or events. Around the same years other studies were conducted on the origin of these gestures and on their role for the emergence of language (for a review see Volterra & Erting, 1990/1994).

The origin of deictic gestures in action was quite evident, in the progression from showing to giving to pointing that demonstrated very clearly a progressive detachment from the object. Only through pointing the child does become able to refer to an object without directly grasping or touching it. Some authors have attributed a special role to pointing. Bruner (1975), for example, describes it as an important way of establishing the joint attention situations within which language will eventually emerge (see also Lock, 1997; Lock, Young, Service, & Chandler, 1990; Masur, 1983. For a recent review see Kita, 2003) .

Specifically these gestures provide the infant with a means of redirecting the attention of another member of the same species and of making reference to things.

Another means for making reference to things is symbolic play (McCune, 1995). Volterra and colleagues (1979) have highlighted interesting parallels in the content and sequence of development (gradual decontextualization) of symbolic play schemes and early word production.

In a subsequent longitudinal diary study of one Italian infant followed from the age of 10 to 20 months, Caselli (1983, 1990) reported that many of the ***actions*** usually set aside as "schemes of symbolic play" (e.g, holding an empty fist to the ear for TELEPHONE) were in fact ***gestures***, frequently used by the child to communicate in a variety of situations and contexts similar to those in which first words were produced. These gestures, characterized as "referential gestures," or "representational gestures" differed from deictic gestures in that they denoted a precise referent and their basic semantic content remained relatively stable across different situations. Other representational gestures, such as conventional gestures, like waving the hand for BYE_BYE, were non object-related. The form and meaning of these gestures seemed to be the result of a particular agreement established in the context of child–adult interaction, while their communicative function appeared to develop within routines similar to those which Bruner (1983) has considered fundamental for the emergence of spoken language.

The communicative use of representational gestures has been confirmed by Zinober and Martlew (1985) and Acredolo and Goodwyn (1988) in studies of larger groups of British and American children. In addition, subsequent analyses of young children's vocabularies suggested that representational gestures account for a large portion of children's early communicative repertoires. Results from a study of 20 Italian children revealed that, at one year of age, these children made extensive use of both the gestural and the vocal modalities in their efforts to communicate, and that it was only in a subsequent phase that the vocal modality became the predominant mode of communication (Caselli, Volterra, Camaioni, & Longobardi, 1993).

In the last decade research conducted by different laboratories began to explore the role of gesture not only in the earliest stage of language development but also in the subsequent stage, during the transition from one to two word utterances (Blake, 2000; Butcher & Goldin-Meadow, 2000; Capirci, Caselli, Iverson, Pizzuto, & Volterra, 2002; Goldin-Meadow, 2002; Goldin-Meadow & Butcher, 2003; for a recent review see Capone & McGregor, 2004).

Data from a study of 12 Italian children, videotaped at home when they were 16 and 20 months of age, suggest that during the first half of the second year, gestures may even account for a larger proportion of children's communicative repertoires and overall production than do words (Iverson, Capirci, & Caselli, 1994). Results indicated that while gestures accounted for a substantial portion of the children's repertoires at both ages, gestures were most prevalent in children's communication at 16 months. By 20 months, a clear shift toward a preference for communication in the vocal modality was observed: the majority of children had more words than gestures at this age.

Just as gestures provide a way for young children to communicate meaning during early lexical acquisition, they also play a transitional role in the development of the ability to convey two pieces of information within a single, communicative utterance. Recent research has examined this issue with regard to developmental changes in the structure of children's utterances.

With regard to the structure of early gestural and vocal utterances, Capirci, Iverson, Pizzuto, & Volterra (1996) reported clear developmental changes in gesture production in single- as compared to two-element utterances produced by the previously described Italians 16 and 20 month olds. In line with findings reported by other researchers (e.g., Butcher & Goldin-Meadow, 2000; Goldin-Meadow & Morford, 1990), they noted that all of the children in their sample produced cross-modal combinations consisting of a single gesture and a single word while they were still one-word speakers. Indeed, at both ages, the most frequent two-element utterances were gesture-word combinations; and production of these combinations increased significantly from 16 to 20 months. In addition, despite the fact that children readily combined gestures with words, combinations of two representational gestures were very rarely observed. When children combined two representational elements, they did so in the vocal modality.

These findings on the role of gesture in the acquisition and development of language were mainly presented and discussed among developmental psychologists or linguistics interested in the topic of language acquisition, but they did not raise a particular interest in a larger audience. In more recent years a new theoretical framework emerging from different disciplines (linguistics, antrophology, neurophysiology) makes this approach to the ontogeny of language extremely relevant.

According to a linguistic perspective gesture is part of language and language itself is considered a gesture-speech integrated system (Kendon, 2004; McNeill, 1992, 2000). Acts of speaking and gesturing are bound to each other in time at a general level. McNeill (1992, 2000) claims that the extremely close synchrony between gesture and speech indicates that the two operate as an inseparable unit, reflecting different semiotic aspects of the cognitive structure that underlies them both.

According to an evolutionary perspective language phylogenetically evolved from a manual system and the most recent formulation of the theory of a gestural origin of language (Corballis, 2002) has proposed that gesture has existed side by side with vocal communication for most of the last two million years, a hypothesis that has also been put forward by other scholars (Hewes, 1976; Armstrong, Stokoe, & Wilcox, 1995). Gesture was not simply replaced by speech, rather, gesture and speech have co-evolved in complex interrelationships throughout their long and changing partnership.

The tight relationship between language and gesture described above is compatible with recent discoveries regarding the shared neural substrates of language and meaningful actions that, in the work developed by Rizzolatti's laboratory (Gallese, Fadiga, Fogassi, & Rizzolatti, 1996; Rizzolatti & Arbib, 1998) have been linked to gestures. Specifically, Rizzolatti and his colleagues have demonstrated that hand and mouth representations overlap in a broad frontal-parietal network called the "mirror neuron system," which is activated during both perception and production of meaningful manual action and mouth movements. These neurons respond both when the monkey makes a grasping movement and when it observes the same movement made by others. Single mirror neurons have only been measured in macaque monkeys. Human brain imaging data only provide evidence for mirror neuron systems. The discovery of "mirror systems" provided a significant support to the notion of a gestural origin of human language and represents the basic mechanism from which language could have evolved (see Armstrong et al., 1995; Corballis, 2002).

Mirror neurons create a direct link between the sender of a message and its receiver. Through them, therefore, observing and doing become manifestations of a single communicative faculty rather than two separate abilities. Its novelty consists in the fact that it indicates a neurophysiological mechanism that may create a common (parity requirement), non-arbitrary, link between the communicating individuals. This link can hardly be created by sounds alone. Sounds, by their nature, cannot generate the shared, non-arbitrary knowledge that can be achieved through the involvement of the motor system.

In the present study we use this theoretical framework to present empirical longitudinal data on the early stages of language development in three Italian children. Our goal is to investigate the relationship between gestures and words during early stages of language acquisition, extending our findings in the period preceding and following the two age points considered in our previous studies, focusing on 16 and 20 months of age. Furthermore, within the framework of the mirror neuron system, we want to determine if meaningful manual actions precede and pave the way to the development of language, and if they share a semantic link with gestures and words.

Method

Participants and procedure

The participants of this study were three typically-developing children (2 second-born boys and 1 first-born girl) videotaped monthly in their homes during a spontaneous play situation when they were between 10 and 23 months of age. Each session lasted approximately 30 minutes, during which the children interacted and played with their mothers. The play sessions were not structured by the experimenter and mothers were encouraged to engage their children in play and conversation as they normally would. The observations were divided equally into three 10-minutes segments so that the children were filmed in three different contexts: play with new examples of familiar objects, play with familiar objects, and a meal or snacktime. The procedure was similar to that adopted by Iverson et al. (1994) and Capirci et al. (1996). The new objects included a set of toys provided by the experimenter: a toy telephone, a plate, a cup, a toy glass, two animal picture books, a spoon, a teddy bear, two small cars, a ball and two combs. Familiar objects varied with each child and included, for example, books, toy cars, toy animals, balloons, and blocks.

Table 1 reports the age range during which each child was observed and the number of sessions conducted during this period. Table 1 also presents the ages at which each child first produced a one-word utterance and a two-word combination (at least two examples).

Table 1. Participants

Name	Sex	Period of data collection	Number of sessions	Age of first word	Age of two-word combination
Luigi	M	10 to 21 months	10	10	21
Marco	M	10 to 23 months	14	11	18
Federica	F	10 to 23 months	13	12	18

One of the children (Luigi) was already producing words during the first observation session (10 months); the remaining two children (Marco and Federica) produced their first words during the observation period respectively at the ages 11 and 12 months. The ages at which the children began producing two-word combinations during sessions were 21 for Luigi, 18 for Marco and Federica.

Coding

All communicative and intelligible actions, gestures and speech produced by children alone or in combinations, were transcribed and coded. Action, gestures and speech were considered to be communicative if they were accompanied by eye contact with another person, vocalization, or other clear evidence of an effort to direct the attention of another person present in the room (Thal & Tobias, 1992).

SPEECH CODING: All the communicative speech produced by each child was coded and classified into one of two categories: Words and vocalizations. Words were utterances that were either actual Italian words (*mamma*, 'mommy') or "baby-words" (words used or pronounced in a manner different from Italian adult usage) that were used consistently to refer to the same referent throughout the observation (*ncuma* for *ancora*, 'more').

Vocalizations were utterances not used consistently to refer to a particular referent but that appeared to be communicative nonetheless. Vocalizations produced alone or in combination with gestures were transcribed but not further analyzed for the present study.

Words were classified as deictic or representational. Deictic words included demonstrative and locative expressions (e.g. 'this', 'there'), and personal and possessive pronouns (e.g., 'I', 'yours'). Like deictic gestures, the precise referent of these words can only be established by making reference to the context in which they are used. Representational words included, for the most, "content words" that in the adult language are classified as common and proper nouns, verbs, adjectives (e.g., 'mommy', 'flowers', 'Luigi', 'open', 'good'), affirmative and negative expressions (e.g., 'yes', 'no', 'all gone'), and also conventional interjections and greetings such as 'bravo!', or 'bye bye'.

GESTURE CODING: All the gestures were transcribed and classified as deictic or representational (all gestures described are denoted in capital letters).

Deictic gestures are those gestures that refer to an object or event by directly touching or indicating the referent. The meaning of these gestures can only be determined through reference to the context in which communication occurs.

Deictic gestures included: SHOW, GIVE, REQUEST, POINT. A gesture was recorded as SHOW when the child held up an object in the adult's line of sight, as GIVE when the child gives an object to the adult. REQUEST was defined as an extension of the arm, sometimes with repeated opening and closing of the hand. Gestures were classified as POINT if there was clear evidence of an extension of the index finger directed toward a specific object, location or event. Following Thal and Tobias (1992), instances of patting a location or object were also coded as pointing.

The criteria for isolating representational gestures were: Manual and/or body movements directed to another individual that were neither direct manipulation of objects nor body adjustments (Ekman & Friesen, 1969; Kendon, 1980). An action to be coded as a gesture required some distance (in time, space, or content) between the movement and that to which it refers; a gestures also requires some instances of intentionality (Blake, 2000; Caselli & Volterra, 1990; Goodwyn & Acredolo, 1993). We excluded all acts made with an object-referent in hand, denoting them as functional object use, play or meaningful action.

Representational gestures included all gestures that referred to an object, person, location, or event through hand movement, body movement, or facial expression. These gestures differ from deictic gestures in that they represent specific referents, and their basic semantic content does not change appreciably with the context. In order to ascertain the stability of the form, gestures were described in terms of the shape of the hand, the type of movement, and the place of articulation.

They included: Gestures iconically related to actions performed by or with the referent (e.g., Bringing empty hand to lips for SPOON; Holding empty fist to the ear for TELEPHONE); gestures describing qualities or characteristics of an object or situation (e.g., extending the arms for BIG or waving the hands for TOO HOT); gestures representing intransitive actions (e.g.: moving rhythmically the body without music for DANCING; covering the eyes with one hand for PEEK-A-BOO); conventional gestures (e.g., shaking the head for NO, turning and raising the palms up for ALL_GONE), including culturally-specific gestures proper to the Italian repertoire (e.g.: bringing the index finger to the cheek and rotating it for GOOD or opening-closing four fingers, thumb extended, for CIAO = 'bye-bye').

ACTION CODING: All communicative and intelligible manual actions associated with specific objects (e.g.: bringing a phone-handset to the ear; pushing a little car) and intransitive actions (e.g.: dancing with the music; hiding himself under the table) were transcribed. We coded the form of each motion, mentioning the object acted on and/or the context.

The distinction between actions and gestures was sometimes difficult to determine. Actions and gestures produced in a communicative context are not clearly separate categories. Rather they should be considered as a continuum and even adults may produce gestures with an object in hand for communicative purposes.

Results

We begin by describing the developmental transition from one element to two element, focusing on the structure of early gestural and spoken utterances. Then, we present the data that demonstrate the link between gestures and words in that period, specifically: the size of children's gestural and spoken repertoires (the number of distinct "lexical" items — types), the frequency of gestures and words production (tokens), and the relationship between different gestural and spoken categories. Finally, we present data that show the relationship between actions and linguistic communication (gestural and vocal).

Structure of early gestural and spoken utterances

The number of different utterances produced by each child during the observation period (gesture-alone, word-alone, gesture-word combination, word-word combination) are presented in Figure 1.

Gesture-gesture combinations are not reported, because they were produced with a very slow frequency by all children (G-G total tokens for L: 4; for M: 4; for F: 8). Multi-element utterances are not reported in the figure because they were produced by all children only in the last sessions. Interestingly the two children, who produced utterances of three or four words at the end of the period examined (22 months for both) had already produced utterances of two-words and one gesture (M: 21 months and F: 20 months).

A similar developmental pattern was noted in all children (see Figure 1). In the first months of observation all children communicated more frequently with gesture alone, but the duration of this period varied (L: from 10 to 14 months; M: from 10 to 16 months; F: from 10 to 14 months).

Single word and gesture-word utterances followed a similar pattern: both appeared around the same period (L:12 months; M:16 months; F:12 months) and both surpassed the production of single gesture utterances around the same session (L:20 months; M:18 months; F:17 months).

The first two word combinations appear for all three children after the emergence of gesture-word combinations (L: 21 months; M: 18 months; F: 18 months) and correspond to an increase in the production of single word and gesture-word utterances.

In the last sessions single word utterances remain the most frequent productions (L:102; M:110; F: 64), the production of gesture-alone utterances is quite infrequent (L:10; M:9; F:4), while the gesture-word combinations are still more frequent than two-word combinations (L: 39 versus 21; M: 49 versus 33; F: 41 versus 29).

Production of words and gestures

In order to provide an accurate picture of word and gesture production (types) and usage (tokens), we report the data in terms of pattern exhibited by individual children.

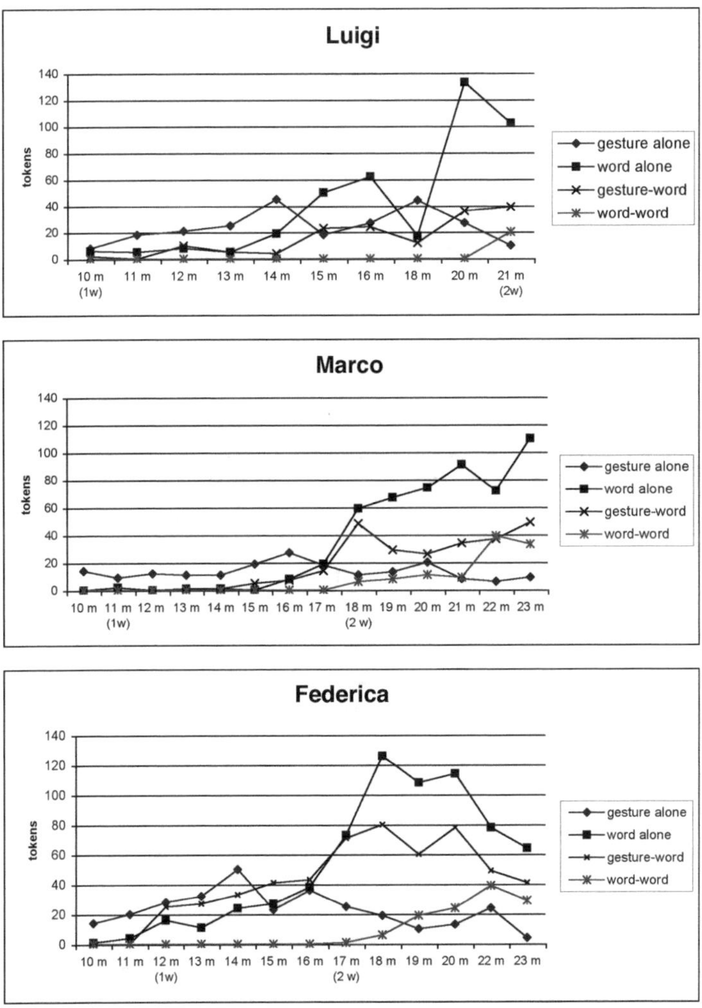

Figure 1. Structure of gestural and spoken utterances

Figure 2 shows the total number of different gesture and word types produced by each child at each session (the appearance of two word utterance is indicated by 2W).

As shown in Figure 2 the three children exhibited a similar developmental pattern: in the first observation sessions they had more extensive gestural than spoken repertoires; afterwards the children appeared to know a similar number of words and gestures; in the final sessions all children had more word than gesture types.

The size of the gestural repertoires looks very similar in the three children and remained relatively stable across all the considered period (gesture types range for L: 7–18; for M: 3–14; for F:8–21). The size of the spoken repertoires, in contrast,

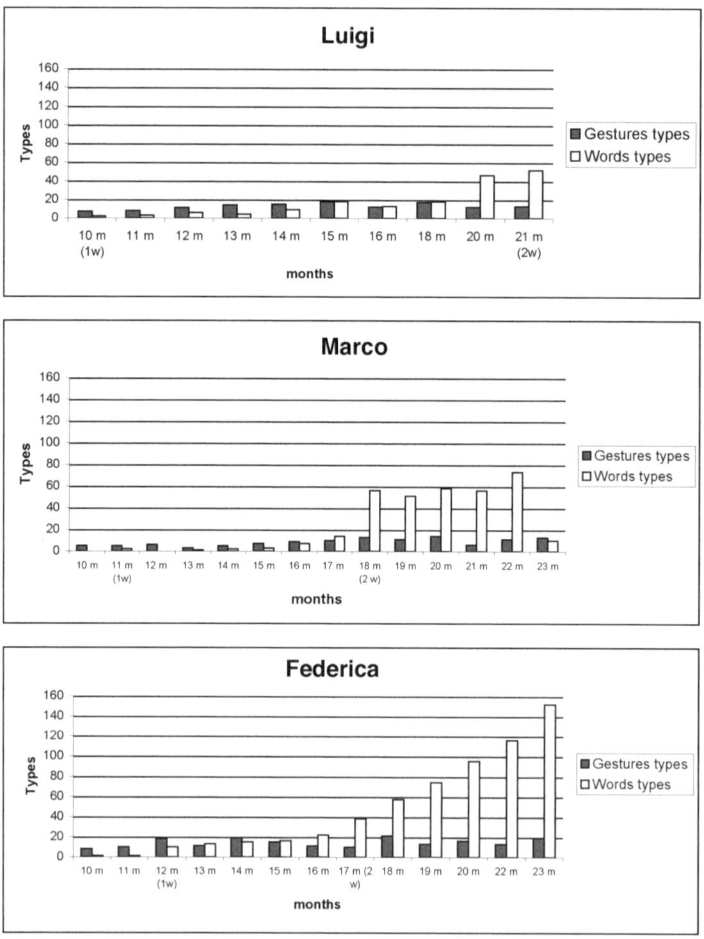

Figure 2. Words and gestures types

increased during the period considered and showed a higher variation across children (word types range for L: 2–52; for M: 0–106; for F: 0–152).

Nevertheless the size of the word repertoire at the emergence of two-word speech was quite similar for all children (L: 52; M: 56; F: 38).

A similar pattern was evident in the production of word and gesture tokens (Figure 3).

The relationship between gestures and words changed over time: at the beginning, children demonstrated a clear preference for gestural communication; in a

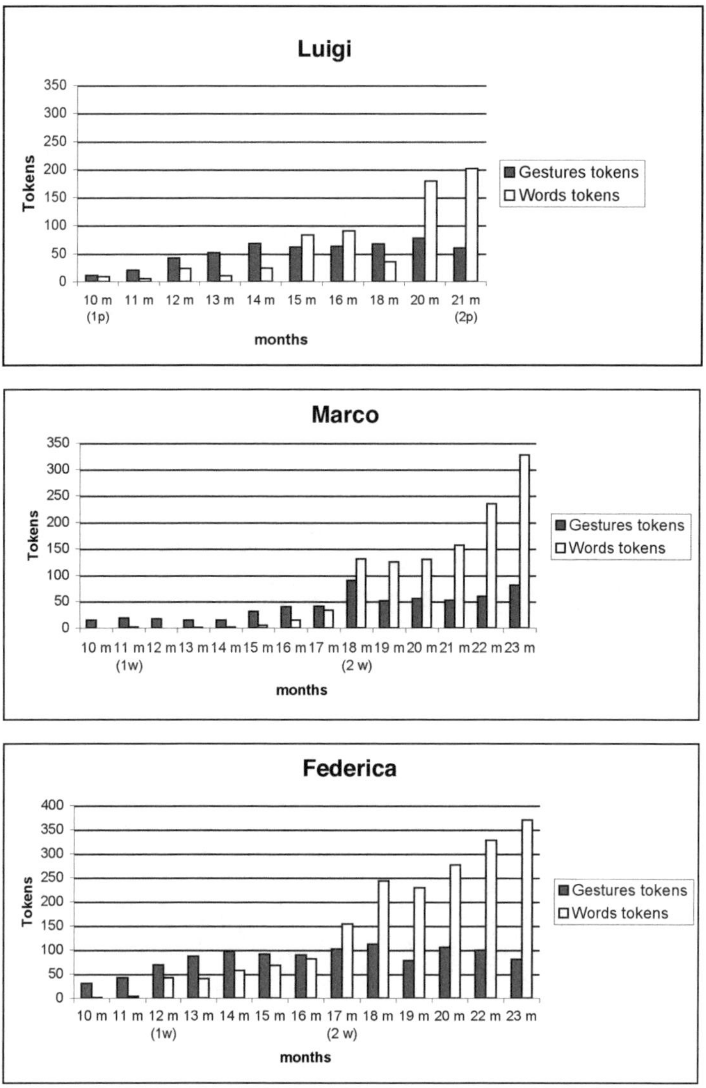

Figure 3. Words and gestures tokens

second period children made an extensive use of both the gestural and the spoken modalities in their efforts to communicate; in the final observation sessions, a clear shift toward the vocal modality was observed.

As shown in both Figure 2 and 3 a clear preference for communication in the spoken modality was observed in the three children just before or at the same time as the emergence of two-word speech (Luigi: 20 months; Marco: 18 months; Federica: 17 months).

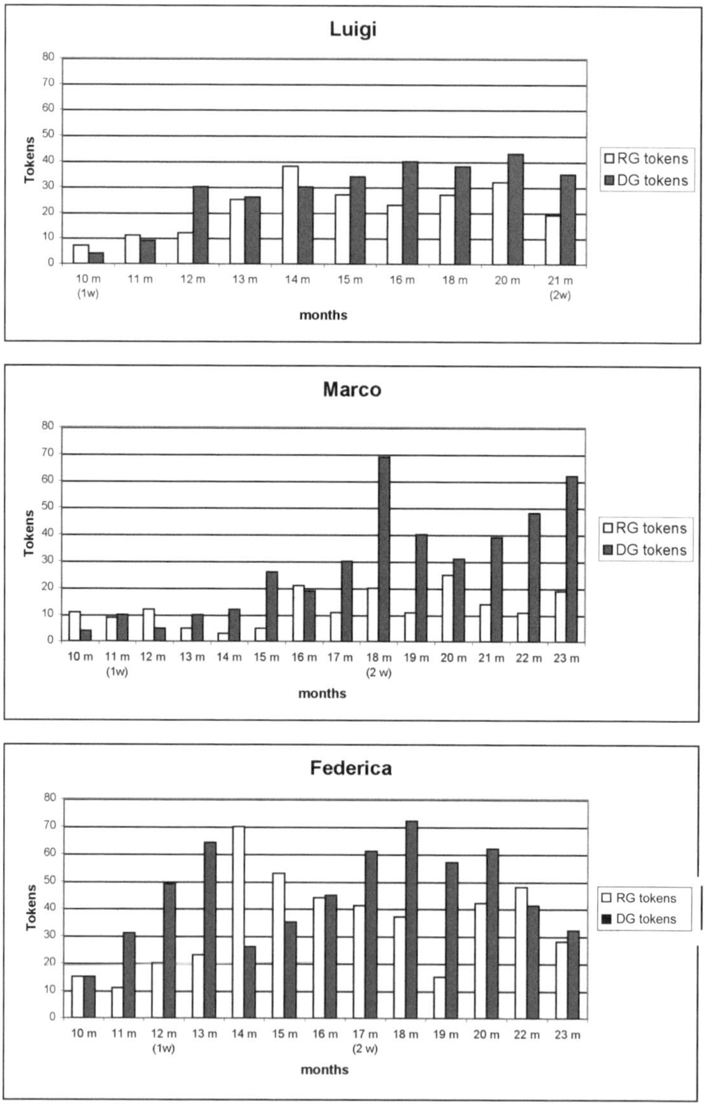

Figure 4. Representational and deictic gestures tokens

Distribution of deictic versus representational elements

Within the spoken and the gestural modalities we further analyzed the types and frequencies of use of the two categories considered: deictic and representational.

Within the gestural modality deictic and representational elements were present from the beginning. The deictic gestural repertoire was restricted to only four gestures (give, show, request and point) that were used frequently during all the period considered (see Figure 4).

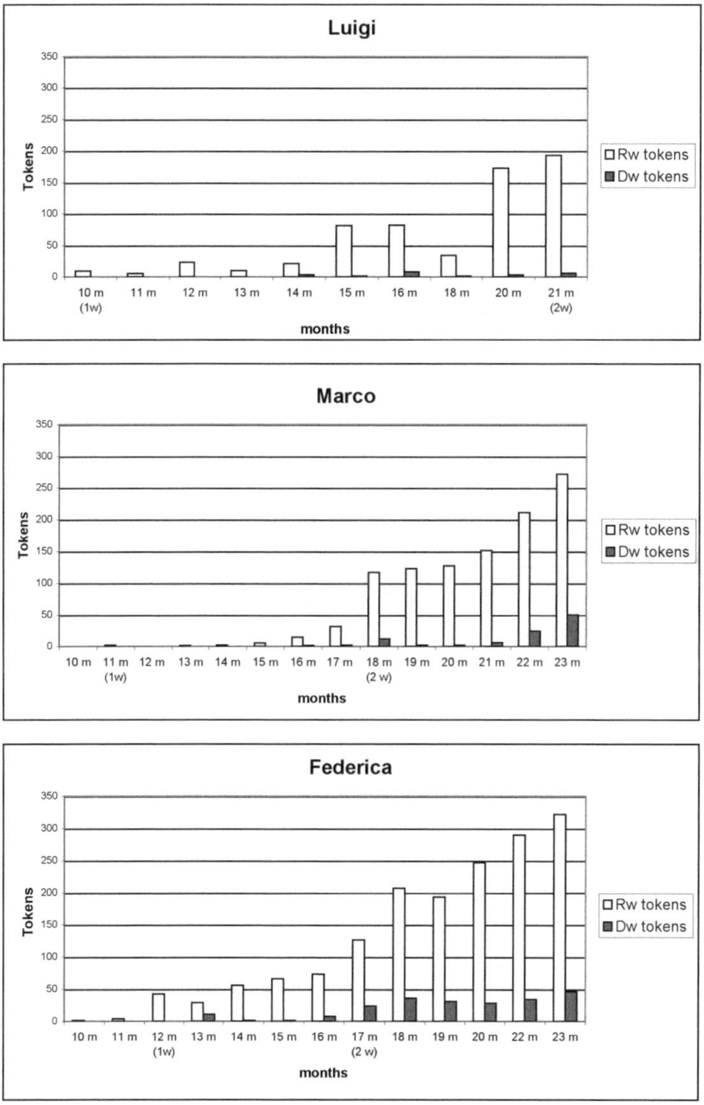

Figure 5. Representational and deictic words tokens

The representational gestural repertoire displayed a higher variation than the deictic gesture repertoire (RG types range for L: 6–14; for M: 1–10; for F: 6–16), but these gestures were used less frequently than the deictic gestures by all children (see Figure 4).

Within the spoken modality only representational elements were present from the beginning, while deictic words appeared later (for M and F with the two-word utterances), and increased in number (dw types range for L.: 0–3; for M.: 0–9; for F.: 0–11) and frequency of use only in the last sessions (see Figure 5). For all three children the first deictic word was "lì" (there) combined with pointing gesture. Representational words showed a much higher variation in types (rw types range for L.: 2–49; for M.: 0–93; for F.: 1–140) and in frequency of use (see Figure 5).

These findings indicated that deictic and representational elements are differently distributed in gestural and in vocal modalities.

The relationship between action and linguistic communication

All meaningful manual and body actions produced by the three children were considered in order to analyze a possible semantic correspondance with gestures and/or words.

In Table 2 a list of examples of actions which overlapped with gestures and/or words in meaning is reported for the three children.

In Figure 6 the percentage of actions performed (produced by each child during all the period considered) that shared the same meaning with a representational gesture and/or a representational word is reported.

Almost all the actions performed were also expressed by representational gestures and/or a representational words. An action could share meaning with a representational word or a representational gesture only, or with both. We observed a total proportion of meaning correspondance of 97,2% for Luigi (Act. = rw: 33.3%; Act. = RG: 30.5%; Act. = RG ***and*** rw: 33.3%), of 88,7% for Marco (Act. = rw: 53.2%; Act. = RG: 0%; Act. = RG ***and*** rw: 35.5%;) and of 97,5% for Federica (Act. = rw: 52.5%; Act. = RG: 0%; Act. = RG ***and*** rw: 45%).

Table 2. Examples of meaning correspondences between actions, gestures and words produced by the three children

Action	Gesture	Word
Bringing an empty spoon to lips	Bringing empty hand to lips	"Pappa" (eat; food)
Bringing a phone-handset to the ear	Holding empty fist to the ear	"Pronto" (hallo)
Pushing a little car	Pushing motion	Brum brum
Dancing with music	Swinging waving the arms	"Ballerina" (dancer)
Blowing out a candle	Blowing out	"Soffi" (you blow)

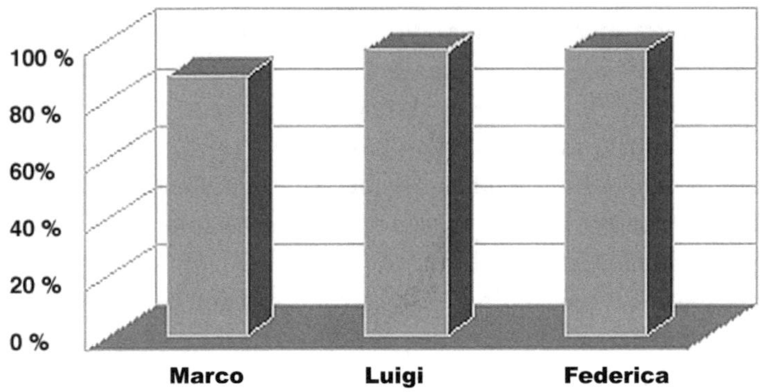

Figure 6. Proportion of semantic correspondance between actions, gestures and/or words

We observed that actions that had a meaning correspondance with gestures and/or words were produced by children before the emergence of the corresponding gesture and/or word (for L: 83%; for M: 91%; for F: 84.6%).

Discussion

We have designed this study to explore two specific aspects of the emergence of language. First, we examined the link between gestures and words in the period from 10 to 23 months in order to confirm the common mechanism underlying both modalities. The second aspect investigated is the possible link between meaningful actions, gestures, and words, in order to ascertain if action may be considered the first step toward the emergence of communication ability (Zukow-Goldring, 2006).

In the present study we found that all three children used gestures during all the period considered to request and/or to label. In particular, the children began to communicate intentionally mainly through gestures. Around 15 –17 months there was a basic "equipotentiality" between gestures and words. This was a bimodal period, as defined by Abrahamsen (2000), in which "words are not as distinct from gestures and gestures are not as distinct from words as they first appear". Both modalities were used productively to communicate about a specific referent in a decontextualized, symbolic manner. At the end of the observed period we noted a shift from symbolic/representational communication in gestural modality to symbolic/representational communication in vocal modality. This shift could not be simply attributed to a contraction of the children's gestural repertoire, but was due to a parallel, and comparatively greater expansion of the vocal repertoire characterized by a marked increase of one-word utterances and gesture-word combinations that mark the transition to the two-word stage. Furthermore, deixis is primary expressed through gestural modality during all the period considered, while a shift is evident in the representational abilities. They are expressed at the beginning through gestural modality, while expressed mainly through vocal modality when the use of words increases (this topic is explored in more details in Pizzuto & Capobianco, 2005).

In addition, all three children produced meaningful communicative actions from the first session, even before they produced the first words. Most of the actions produced by them had a "meaning correspondence" with gestures and/or words that were produced later showing that the emergence of a particular action preceded the production of the gesture and/or word with the corresponding meaning. The meanings shared through the goal directed action were almost all expressed later in a symbolic way with gestures and words.

Taken together the data presented and discussed in this paper support and extend previous studies, providing data preceding and following the two age points previously considered (16 and 20 months of age: Iverson et al., 1994; Capirci et al., 1996). In the present study we were able to show that previous to 16 months the children communicate mainly with gesture alone utterances, and their cross-modal utterances precede in all children the emergence of two-word utterances. After 20 months, multi-element crossmodal utterances (two-words and one gesture) precede in all children utterances of three or four words, and could be considered preparatory for the longer spoken unimodal utterances.

Furthermore our findings provide support for the phylogenetic and neurophysiological claims that there is a tight relationship between the gestural and vocal modalities. This link provides the basis for a developmental model of language in human ontogeny that goes from action to gesture and speech .

We noted a similar developmental pattern in the relationship between gestures and words in all of the three children that seems to mirror the evolutionary scheme proposed by Corballis (2002) in which both modalities co-evolved in a complex interrelationship throughout their long and changing partnership.

Corballis' evolutionary views on a slow transition from gesture to vocal language appear to be supported by our developmental data, as this transition, and the interdependence between gesture and speech, are still evident in children's communicative and linguistic development. As observed by Deacon, it is of course unlikely that language development recapitulates "language evolution in most respects (because neither immature brains nor children's partial mapping of adult modern languages are comparable to mature brains and adult languages of any ancestor)" (Deacon, 1997, p. 354), but we can gain useful insights into the organization and evolution of both language and gesture by investigating the interplay between these modalities in the communication and language systems of children.

The tight relationship between gesture and word may, indeed, be related to action because of the representational property of the motor system (Rizzolatti, Fadiga, Gallese, & Fogassi, 1996; Gallese, 2000). The finding of "mirror" properties in the motor neurons has been the starting point toward finding a neural link between language and motor functions. The parity property of language, that is, what counts for the speaker must count approximately the same for the hearer, is manifested in the action because of the function of the "mirror system" to link self-generated actions and the similar actions of others (Rizzolatti & Arbib, 1998). It has been speculated that mirror systems, in monkeys and in infants, are not restricted to recognition of an innate set of actions but can be recruited to recognize and encode an expanding repertoire of novel actions. This proposal was based on the consideration that mirror systems create a direct link between the sender of a message and its receiver. Through them, therefore, observing and doing become manifestations of a single communicative faculty rather than two separate abilities. The recent finding that audio-visual mirror neurons can be activated by the sound that co-occurs with the action (Kohler et al., 2002) provides further support for this hypothesis providing a possible neural basis underlying this step in the communicative development.

In conclusion, our findings show that gesture and speech are linked to, and co-evolve in, the ontogeny of language. In this process of developing communication gesture and action play a basic function because they have two prerequisites for the emergence of vocal language: both are crucial for attention sharing and meaning sharing. For the first time we are able to show a link not only between "actions and words" or "gestures and words", but also a progression from action to language through gesture.There is suggestive evidence that mirror systems may be related to the functions of attention and meaning sharing and suggest an evolutionary path from action to language.

At least two interesting questions still remain open: the role of caregiver in guiding children behaviour, and the origin of social/interactive gestures.

In our study we considered the children's behaviour. We did not analyze the adult behaviour even though we transcribed adult productions relevant in the interaction.

It is likely that the transition from an object-related mirror-neuron system to a truly communicative mirror-neuron system is related to the development of imitation (see Arbib, 2005). This enrichment of the mirror-neuron system probabily did not evolve originally in order to communicate, but was a consequence of the necessity to learn, by imitation, actions done by others. The necessity to keep track of precise movements sharpened the mirror system and its capacity to convey information. In learning novel actions, a basic function is played by caregiver who guides the infant towards integrating perception and action, the two side of the mirror system (Zukow-Goldring, 2006).

Preliminary observation of our data confirm the relevant role played by caregivers. Almost all actions were produced by the three children in a situation in which the caregiver was present and was making comments and attributing meaning to the action performed by the child.

In future work we would like to analyze in more detail the ways in which imitation, especially assisted imitation, contributes to communicative development, and to test experimentally if caregiver assistance can speed up learning and communicative abilities.

With regard to the second question we were able to show that almost all of the meanings expressed through transitive and intransitive actions were expressed later by the children through gestures and/or words. It is also true, however, that other gestures and words do not seem to originate from object manipulation. For example the gesture/word "no", the gesture/word "ciao" or the gesture "clapping hands" (with "bravo" as corresponding word) begin inside routines and exchanges with the caregivers which seem to have as main goal the communication itself. In these examples the role of assisted imitation provided by caregiver is even more clear. But in these cases the gesture and/or word does not originate from object manipulation plus social interaction but from social interaction exclusively.

In the future we would like to continue to design studies and collect data on the early stages of communicative and linguistic development in order to provide further empirical evidence of the manner in which language emerges from action, reflecting a neural function shared by the motor and the linguistic system.

Acknowledgements

This work, as part of the European Science Foundation EUROCORES Programme OMLL, was supported by funds from the Italian National Research Council and the EC Sixth Framework Programme under Contract no. ERAS-CT-2003-980409.

We thank Donna Thal for her helpful comments on the paper and to Lisa di Lemma for her final editing assistance. We also want to thank the children who participated in the study and their parents.

References

Abrahamsen, Adele (2000). Explorations of enhanced gestural input to children in the bimodal period. In Karen Emmorey & Harlan Lane (Eds), *The signs of language revisited: An antology to honor Ursula Bellugi and Edward Klima* (pp. 357–399). Mahwak: Erlbaum.

Acredolo, Linda & Susan Goodwyn (1988). Symbolic gesturing in normal infants. *Child Development, 59*, 450–466.

Arbib, Michel A. (2005). From monkey-like action recognition to human language: An evolutionary framework for neurolinguistics. *Behavioral and Brain Sciences, 28*, 105–124.

Armstrong, David F., Williams C. Stokoe, & Sherman E. Wilcox (1995). *Gesture and the nature of language*. Cambridge University Press.

Bates, Elizabeth, Laura Benigni, Inge Bretherton, Luigia Camaioni, & Virginia Volterra (1979). *The emergence of symbols: Cognition and communication in infancy.* New York: Academic Press.

Bates, Elizabeth, Luigia Camaioni, & Virginia Volterra (1975). The acquisition of performatives prior to speech. *Merril Palmer Quarterly, 21* (3), 205–226.

Blake, Joanna (2000). *Roots to language: Evolutionary and developmental precursors.* Cambridge, UK: Cambridge University Press.

Bruner, Jerome (1975). The ontogenesis of speech acts. *Journal of Child Language, 2*, 1–19.

Bruner, Jerome (1983). *Child's talk: Learning to use language*. New York: Norton.

Butcher, Cynthia & Susan Goldin-Meadow (2000). Gesture and the transition from one- to two-word speech: When hand and mouth come together. In David McNeill (Ed.), *Language and gesture* (pp. 235–257). Cambridge: Cambridge University Press.

Capirci, Olga, Maria Cristina Caselli, Jana M. Iverson, Elena Pizzuto, & Virginia Volterra (2002). Gesture and the nature of language in infancy: The role of gesture as transitional device enroute to two-word speech. In David F. Armstrong, Michael A. Karchmer, & John V. Van Cleeve (Eds.) *The study of Sign Languages — Essays in honor of William C. Stokoe* (pp. 213–246). Washington, D.C.: Gallaudet University Press.

Capirci, Olga, Jana M. Iverson, Elena Pizzuto, & Virginia Volterra (1996). Gestures and words during the transition to two-word speech. *Journal of Child language, 23*, 645- 673.

Capone, Nina C. & Karla K. McGregor (2004). Gesture development: A review for clinical and research practices. *Journal of Speech Language and Hearing Research, 47*, 173–186.

Caselli, Maria Cristina (1983). Gesti comunicativi e prime parole. *Età Evolutiva, 16*, 36–51.

Caselli, Maria Cristina (1990). Communicative gestures and first words. In Virginia Volterra & Carol J. Erting (Eds.), *From gesture to language in hearing and deaf children* (pp. 56–67). Berlin / New York: Springer-Verlag. (1994 — 2nd Edition Washington, D.C.: Gallaudet University Press).

Caselli, Maria Cristina & Virginia Volterra (1990). From communication to language in hearing and deaf children. In Virginia Volterra & Carol J. Erting (Eds.), *From gesture to language in hearing and deaf children* (pp. 263–277). Berlin / New York: Springer Verlag. (1994 — 2nd Edition Washington, D.C.: Gallaudet University Press).

Caselli, Maria Cristina, Virginia Volterra, Luigia Camaioni, & Emiddia Longobardi (1993). Sviluppo gestuale e vocale nei primi due anni di vita. *Psicologia Italiana*, IV, 62–67.

Corballis, Michael C. (2002). *From hand to mouth — The origins of language*. Princeton, NJ: Princeton University Press.

Deacon, Terence (1997). *The symbolic species. The coevolution of language and the human brain.* London: The Penguin Press.
Ekman, Paul & Wallace Friesen (1969). The repertoire of non-verbal behavior: Categories, origins, usage and coding. *Semiotica 1(1)*, 49–98.
Gallese, Vittorio (2000). The inner sense of action: Agency and motor representations. *Journal of Consciousness Studies, 7*, 23–40.
Gallese, Vittorio, Luciano Fadiga, Leonardo Fogassi, & Giacomo Rizzolatti (1996). Action recognition in the premotor cortex. *Brain, 119*, 593–609.
Goodwyn, Susan W. & Linda P. Acredolo (1993). Symbolic gesture versus word: Is there a modality advantage for onset of symbol use? *Child Development, 64*, 688–701.
Goldin-Meadow, Susan (2002). *Hearing gestures: How our hands help us think.* Cambridge, MA: Harvard University Press.
Goldin-Meadow, Susan & Cynthia Butcher (2003). Pointing toward two-word speech in young children. In Sotaro Kita (Ed.), *Pointing. Where language, culture, and cognition meet* (pp. 85–107). London: Lawrence Erlbaum Associates.
Goldin-Meadow, Susan & Marylin Morford (1990). Gesture in early child language. In Virginia Volterra & Carol J. Erting (Eds.), *From gesture to language in hearing and deaf children* (pp. 249–262). Berlin / New York: Springer-Verlag. (1994 — 2nd Edition Washington, D.C.: Gallaudet University Press).
Hewes, Gordon W. (1976). The current status of the gestural theory of language origin. *Annals of the New York Academy of Sciences*, vol. 280, 482–604.
Iverson, Jana M., Olga Capirci, & Maria Cristina Caselli (1994). From communication to language in two modalities. *Cognitive Development*, 9, 23–43.
Kita, Sotaro (Ed.) (2003). *Pointing. Where language, culture, and cognition meet.* London: Lawrence Erlbaum Associates.
Kendon, Adam (1980). Gesticulation and speech: Two aspects of the process of utterance. In Mary R. Key (Ed.), *The relationship of verbal and nonverbal communication* (pp. 207–227). The Hague: Mouton and Co.
Kendon, Adam (2004). *Gesture. Visible action as utterance.* Cambridge: Cambridge University Press.
Kohler, Evelyne, Christian Keysers, Maria Alessandra Umiltà, Leonardo Fogassi, Vittorio Gallese, & Giacomo Rizzolatti (2002). Hearing sounds, understanding actions: Action representation in mirror neurons. *Science, 297*, 846–8.
Lock, Andrew (1997). The role of gesture in the establishment of symbolic abilities: Continuities and discontinuities in early language development. *Evolution of Communication, 1(2)*, 159–193.
Lock, Andrew, Andrew Young, Valerie Service, & Paul Chandler (1990). Some observations on the origins of the pointing gesture. In Virginia Volterra & Carol J. Erting (Eds.), *From gesture to language in hearing and deaf children* (pp. 42–55). Berlin / New York: Springer-Verlag. (1994 — 2nd Edition Washington, D.C.: Gallaudet University Press).
Masur, Elise Frank (1983). Gestural development, dual-directional signaling, and the transition to words. *Journal of Psycholinguistic Research, 12*, 93–109.
McCune, Lorraine (1995). A normative study of representational play at the transition to language. *Developmental Psychology, 31*, 198–206.
McNeill, David (1992). *Hand and mind — What gestures reveal about thought.* Chicago: University of Chicago Press.

McNeill, David (Ed.) (2000). *Language and gesture*. Cambridge: Cambridge University Press.
Pizzuto, Elena & Micaela Capobianco (2005). The link and differences between deixis and symbols in children's early gestural-vocal system. *Gesture*, 5 (1/2), 179–199. (This volume)
Rizzolatti, Giacomo, Luciano Fadiga, Vittorio Gallese, & Leonardo Fogassi (1996). Premotor cortex and the recognition of motor actions. *Cognitive Brain Research, 3,* 131–141.
Rizzolatti, Giacomo & Michel A. Arbib (1998). Language within our grasp. *TINS*, *21*, 188–194.
Thal, Donna & Stacy Tobias (1992). Communicative gestures in children with delayed onset of oral expressive vocabulary. *Journal of Speech and Hearing Research 35*, 1281–1289.
Volterra, Virginia, Elizabeth Bates, Laura Benigni, Inge Bretherton, & Luigia Camaioni (1979). First words in language and action: A qualitative look. In Elizabeth Bates (Ed.), *The emergence of symbols: Cognition and communication in infancy* (pp. 141–222). New York: Academic Press.
Volterra, Virginia & Carol J. Erting (Eds) (1990). *From gesture to language in hearing and deaf children*. Berlin / New York: Springer Verlag. (1994 — 2nd Edition Washington, D.C.: Gallaudet University Press).
Zinober, Brenda & Margaret Martlew (1985). Developmental changes in four types of gesture in relation to acts and vocalizations from 10 to 21 months. *British Journal of Developmental Psychology, 3*, 293–306.
Zukow-Goldring, Patricia (2006). Assisted imitation: Affordances, effectivities, and the mirror system in early language development. In Michael A. Arbib (Ed.), *Action to language via the mirror neuron system*. Cambridge: Cambridge University Press.

The link and differences between deixis and symbols in children's early gestural-vocal system

Elena Pizzuto[1] and Micaela Capobianco[2]
[1]Institute of Cognitive Sciences and Technologies, CNR, Rome /
[2]University of Rome "La Sapienza"

This study aims to contribute to a clearer understanding of children's developing gesture-speech system examining the interrelation between deictic and representational elements of gestural and vocal types. We analyze the spontaneous productions of six children, observed longitudinally from 12 to 24 months during the transition from one- to two- and multi-element vocal utterances. We focus on children's gestural and vocal repertoires, and one- and two-element utterances encoding different information within and across modalities. Results indicate that deictic and representational elements are unevenly distributed in the gestural vs. the vocal modality, and in one vs. two-element utterances, with patterns that differ from those observed in the adult gesture-speech system. In these early stages speech and gesture are interrelated primarily through deictic gestures, and representational abilities appear to be markedly more constrained in the gestural as compared to the vocal modality.

Introduction

The purpose of this paper is to contribute to a more comprehensive understanding of the interrelation between deictic and content-loaded elements of different kinds which, for short, we label here "representational", in children's early gestural-vocal system. We focus on a key, universal phase of language development that occurs in the second year of life: the transition from one- to two- and multi-element utterances, typically accompanied by a significant vocabulary growth (Slobin, 1985; Clark, 2003).

A distinctive feature of deictic elements, whether they are words (e.g. demonstratives or locatives such as "this", "there") or gestures (e.g. the prototypical POINT[1], produced with the index finger extended to direct someone's attention toward something in the environment), is that their interpretation heavily or entirely depends upon contextual information. The referent of a POINT, much as the referent of a word like "this" can be identified only inspecting the physical context of utterance (e.g. looking in the direction being pointed), or also the broader communicative-linguistic context (see for example McNeill's (1992) observations on abstract pointing in narrative texts, or Lyons' (1977) observations on textual deixis).

Although deictic elements are of crucial importance for human communication and language (see among others Kita, 2003; Lyons, 1977), they clearly cannot and should not be easily assimilated to content-loaded, representational elements. Words such as 'dog', 'no' or even onomatopoeic and sound symbolism forms that are frequent in early speech (e.g. 'brum-brum' for 'car'), or more or less conventional or iconic gestures (e.g. shaking the head for NO, opening and closing the hand, mimicking a flashing motion, to refer to a 'FLASHING-LIGHT') convey information that differs in kind from that of deictic elements. Regardless of their arbitrariness (in the case of most words) or iconicity (in the case of most gestures and also of onomatopoeic and sound symbolism forms), these content-loaded elements can only be interpreted referring to symbolic conventions that must be shared by the word or gesture producer and his/her interlocutor.

In the last thirty years, beginning with the pioneering studies described by Bates, Camaioni, and Volterra (1975), Bates (1976), Bates et al. (1979), Greenfield and Smith (1976), a wealth of knowledge has accumulated on the important role that gestures play in early development. Even though differences in data sets, methodology and terminology render often difficult to compare the findings of different studies, there is clear evidence that both deictic and representational gestures of the sorts sketchily described above are present in children's early expressive repertoires and productively used in children's early utterances (for overviews from different perspectives see Blake, 2000; Butterworth, 2003; Lock, 1980; Volterra & Erting, 1990; see also Guidetti, 2002 for critical observations on the need to distinguish iconic from more conventional gestures).

Different studies have explored the role of gestures during the transition from one- to two- and multi-element utterances. Capirci, Iverson, Pizzuto, & Volterra (1996) have reported the existence of a significant relationship between single-gesture and gesture-word, or crossmodal utterances and total vocal production in a group of children observed at 16 and 20 months of age. Longitudinal studies described by Butcher and Goldin-Meadow (2000) and Goldin-Meadow and Butcher (2003) have highlighted significant correlations between crossmodal utterances in which the combined elements appear to encode different meanings, and the onset of two-word utterances.

An appropriate discussion of the studies that have been conducted is beyond the scope of this paper (see Capone & McGregor, 2004, for a recent insightful review; Volterra, Caselli, Capirci, & Pizzuto, 2005, for relevant observations). We wish to focus here on some aspects of children's early gestural vs. vocal expressive abilities which, in our view, have not been sufficiently clarified. These regard the different import that deictic vs. representational elements of gestural vs. vocal types have in structuring children's early repertoires and utterance-patterns, and the extent to which children's early use of representational gestures is, or is not comparable to that noted in adults or older children.

Clarifications on the typology of gestures *and* words used early in development are especially needed, in our view, in the light of the increasing evidence on the inherently multimodal features of adult speech (Kendon, 1996; McNeill, 1992, 2000; Kita, 2003). We refer here in particular to McNeill's (1992) work showing that adults (and also children, from age 4–5 onward) use different types of idiosyncratic, content-loaded gestures (described primarily as "iconics" and "metaphorics") very productively. These gestures are meaningfully and temporally integrated with speech to articulate meaning in different ways across the vocal and gestural modality. It is interesting to note that in adult coverbal gesturing deictic gestures appear to be in markedly smaller proportions compared to representational gestures (McNeill, 1992, p. 93).

The developmental literature provides somewhat different indications. Capirci et al. (1996), Pizzuto (2002) and Volterra et al. (2005), drawing on analyses of the spontaneous production of twelve children examined at 16 and 20 months of age, have found that the gestures children use most productively in their crossmodal combinations are not representational but deictics, most notably a POINT combined with a representational word. This finding is in agreement with what is reported, but often not explicitly discussed, in many recent and less recent studies (e.g. Goldin-Meadow & Butcher, 2003; Greenfield & Smith, 1976).

Capirci et al. (1996), Pizzuto (2002) and Volterra et al. (2005) have proposed that these crossmodal combinations are broadly comparable to forms of nomination (e.g. POINT to flowers while saying 'flowers'), and predication (e.g. POINT to toy while saying 'nice'). As remarked by Pizzuto, Capobianco & Devescovi (2005) in a subsequent longitudinal study, which is closely related to the present one, this widespread use of deictic gestures appears to be peculiar of the early stages of development.

Relevant differences in the gesture types used by children vs. adults have also been highlighted by Mayberry and Nicoladis (2000) who have proposed a more strict definition of meaningful gestures than it is usually adopted in the developmental literature, and compared the use of deictics with that of iconics (gestures depicting actions or objects), and beats (gestures marking discourse structure). Mayberry and Nicoladis' study on five bilingual children observed from 2 to 3 1/2 years of age shows that in this age range the use of iconic and beat gestures correlates with language development, while the use of pointing gestures does not.

It must be recalled that children's early vocal only utterances, unlike their crossmodal utterances, most frequently consist of two representational elements (e.g. "baby highchair", "mommy eggnog" — see Brown's (1973) detailed analyses; Capirci et al. (1996), Volterra et al. (2005), Pizzuto, Capobianco, and Devescovi (2005), for contrastive evidence with crossmodal utterances). This pattern is found also in children acquiring signed languages (see Newport & Meier, 1985; Volterra & Iverson, 1995; Volterra et al., 2005). The significance of this combinatorial pattern cannot be underestimated in the light of Deacon's (1997, pp. 69–100) arguments on the tight links between, on one hand, indexical and symbolic relationships in symbol learning and, on the other hand, the ability to combine two content-loaded elements and the development of truly symbolic abilities, in chimps as well as children.

Taken together, the observations summarized above indicate asymmetries in children's early gestural vs. vocal expressive abilities, and differences between the child and the adult gesture-speech system that can be relevant to reach a clearer understanding of children's developing symbolic system.

The present study stems from the work reported by Capirci et al. (1996), Iverson, Capirci, and Caselli (1994), Pizzuto (2002), Volterra et al. (2005), and constitutes an expansion of Pizzuto et al.'s (2005) longitudinal study. Pizzuto et al. (2005) analysed the development of deictic and representational gestures, vocalizations and words in the spontaneous production of four children, observed longitudinally from 10–12 to 24–25 months of age. Results indicated that while gestural deixis plays a primary role in children's early production, representational abilities are markedly more constrained in the gestural as compared to the vocal modality. In the present study we explore the generalizability of these findings examining a larger set of longitudinal data. We focus on children's one- and two-element utterances, and we extend the analysis to the repertoires of gestures and words (which was not investigated by Pizzuto et al., 2005). We address three interrelated questions:

1. Are deictic vs. representational elements evenly or unevenly distributed in children's gestural and vocal repertoires, and in children's gestural, vocal, or gestural-vocal utterances ?
2. Are the patterns observable similar or different in one- as compared to two-element utterances, and how do they change with development?
3. What do the patterns observable reveal on children's developing gesture-speech system?

Method

Participants and data set

Data were taken from a broader, ongoing longitudinal study of nine typically developing children undertaken by the second author of the present paper as part of her doctoral dissertation (Capobianco, 2006). All children were videotaped at home, between 10–12 and 24–25 months, in spontaneous interactions with their mothers (occasionally also with their father/other caregivers). Observations sessions (lasting 45' on average) were planned monthly, but this schedule could not always be followed and some sessions were skipped. For the present study we selected data on six children for whom data collection has been completed: four were males, two females, four first-born, two second-born and one third born. We chose seven one-month samples for each child at age-points for which we had records that overlapped almost completely for all six children (see Table 1). Three of the children examined in the present study (ALE, GAL and MAR) participated also in Pizzuto et al.'s (2005) study.

Coding and analysis

Children's communicative gestural and vocal productions were identified, classified and coded following the criteria and coding scheme described in Iverson et al. (1994) and Capirci et al. (1996), with modifications concerning the analysis of bimodal utterances and gestures accompanied by vocalizations (see Pizzuto et al., 2005, and below).

All of the children's gestures and words were distinguished in two major classes: deictic and representational. Deictic gestures (DG) included three gestures extensively described in the literature (e.g. Bates et al., 1979): – REQUEST (arm and hand extended, sometimes with a repeated opening/closing of the hand); – SHOW (holding up an object in the adult's line of sight); – POINT (index finger extended).

Table 1. The children examined and their age (in months) at each observation point

Children	Age						
ALE	12	15	16	18	20	23	24
GAL	12	15	16	18	20	23	24
GIA	12	15	17	18	20	22	24
MAR	12	15	16	18	20	23	24
NIC	12	15	16	18	20	23	24
SAV	12	15	16	18	20	23	24

Deictic words (dw) included demonstratives, locatives, person and possessive pronouns and adjectives, as defined in linguistic terms (Lyons, 1977).

Representational gestures (RG) included a heterogeneous set of: – conventional, fairly stereotyped gestures (e.g. shaking the head for NO; nodding for YES, turning and raising up the palms for ALL_GONE, clapping hands for BRAVO!, index finger upward, touching the lips, for SILENCE); – gestures specific of Italian culture (e.g. opening-closing four fingers, thumb extended, for BYE-BYE, touching the cheek with the tip of the index finger, with a rotating movement for GOOD); – iconic facial displays (e.g. a repeatedly opening and closing the mouth for FISH), or arm/hand movements (e.g. flapping the hands for BIRD, moving downward the arm and hand, with an inward motion towards oneself, for COME), or large body movements (e.g. mimicking a horse-riding action, rocking the body back and forth, for HORSE).

Representational words (rw) included: – content words that, in the adult language, are assigned to the class of common and proper nouns, adjectives, verbs, adverbs (e.g. the Italian words for 'mommy, car, small, look'), affirmative and negative expressions (e.g. 'yes', 'no', all_gone'), interjections (e.g. 'bravo!'), greetings (e.g. 'ciao' = bye_bye), adverbials and prepositions such as "up", "down"; – onomatopoeic and sound symbolism forms (e.g. 'brum-brum' for 'car', 'grrr' for 'lion', 'coco" for 'chicken', 'toc-toc' for 'knocking').

As remarked elsewhere (Pizzuto et al., 2005), we recognize that the content-loaded elements we label as "representational" may have an uncertain symbolic status. Mayberry & Nicoladis (2000) have raised serious criticism with respect to the "overly broad" definition of gestures that is adopted in most developmental studies, including the present one. More precise typological and/or functional distinctions among children's early gestures certainly need to be drawn, especially for drawing accurate comparisons with adult gestures (see Guidetti, 2002, for relevant remarks).

However, as noted by Volterra et al. (2005), similar problems arise in the classification of children's early words or word-like sounds. Many of these have an equally precarious symbolic status: a "grrr" sound reliably interpreted, in context, as a label for a pictured lion is no more nor less symbolic than a FISH gesture (repeated opening-closing of the mouth) made to label a pictured fish. The rationale underlying our classification is that the distinction between deictic and representational elements, if made with the same criteria for both gestural and vocal behaviors, can help us uncover salient features of children's developing gesture-speech system which would remain unnoticed if this distinction is not made.

In the analysis of children's utterances, we classified as one-gesture or one-word all gestures and words (as defined above) that occurred in isolation, further distinguishing the type of gesture or word used (1DG, 1RG, 1dw, 1rw).

Gestures and words *or* vocalizations that appeared to be related in terms of information encoded (see below) and/or also temporal organization patterns were classified as two- or multi-element utterances. Within the limits of this paper we do not consider multi-element utterances or the temporal relationship between gestural and vocal elements (see Pizzuto et al., 2005, for relevant evidence on these topics).

Two-element utterances were distinguished, on the basis of the information vehiculated, in three major types: *bimodal equivalent*, *complementary*, and *supplementary.*

Bimodal equivalent utterances included combinations of a gesture with a word *or* a vocalization in which the elements combined vehiculated essentially the same information. Two subtypes were distinguished: (a) combinations of two representational units (RG=rw, where the notation "=" denotes the comparable information), already described by Capirci et al. (1996) (e.g. NO = no; SILENCE=zitto <be-quiet>); (b) combinations of deictic gesture with vocalizations that were highly variable in form but which appeared to fulfill a primary deictic function (DG=dv: POINT [to pictures in a book]=ah!). As noted in Pizzuto et al. (2005) classifying separately this kind of utterances — notoriously very frequent in early communication (e.g. Bates, 1976; Butterworth, 2003) — appears to be useful for providing a more comprehensive view of the interrelationship between gestures and vocal elements at large. Towards this end we also coded separately combinations of RG with vocalizations (RG-v) that played a generic expressive function (e.g. YES-hmm).

Complementary and *Supplementary* utterances included vocal, gestural, and gestural-vocal or crossmodal combinations in which each of the elements combined contributed to specify or articulate the information provided by the other element. The distinction between complementary and supplementary utterances was inspired by earlier work conducted by Goldin-Meadow and Morford (1985, 1990) on gesture-word combinations. However in this study, as in other studies conducted within our team (e.g. Capirci et al.; 1996, Volterra et al., 2005), we extended the use of these terms to the classification of vocal, gestural, and crossmodal utterances and we attribute to them a different meaning, as explained below (see Pizzuto et al., 2005, for relevant observations on these methodological questions).

Complementary utterances typically referred to a single referent, but had one distinctive feature, denoted by an ampersand (&) between the two combined elements: they always included one deictic element (gestural *or* vocal) which provided non-redundant information, singling out or disambiguating the referent specified by a co-occurring representational element (e.g. DG&rw: POINT [to glass of water] & acqua <water>; dw&rw: eccolo & Titti <here_it_is & Tweety>), or by another deictic element (e.g. POINT [to keys] & quette <these>).

Supplementary utterances referred to either (more often) a single, or to two referents, but in all cases each of the combined elements added information to the other one, a feature signaled in our notation by a plus mark (+) between the two combined elements. Examples of such utterances included combinations of: – two representational words (rw+rw, e.g. bosa <=borsa: bag> + beia <=bella: beautiful>); – a deictic and a representational word (dw+rw: eccio <=questo: this> + mamma <mommy>, meaning 'this belongs to mommy'); – a deictic gesture and a representational word (DG+rw: POINT [to toy] + bello 'beautiful'); – a representational gesture and a representational word (RG+rw: ALL_GONE + coco' <=chicken>).

It should be noted that, on the basis of the information vehiculated, only complementary and supplementary utterances can be considered true combinations of two elements, in which each element provides a distinct piece of information (with a different value depending upon the subtype of combination involved). In contrast, and in spite of the two elements implicated, bimodal equivalent utterances can be assimilated to one-element utterances in which a gesture and a word (or a vocalization) are superimposed (most often simultaneously), rather than combined. The same piece of information is given twice, 'reinforcing' or 'magnifying' the intended message (see Capirci et al., 1996; Volterra et al., 2005).

Results and discussion

We describe the developmental patters we found for the children as a group noting, whenever relevant, the cases in which individual developmental profiles differed significantly from those described for the entire group.

Repertoire

Figure 1 shows the mean number of distinct DG, RG, rw and dw (types) identified in the children's repertoire at each observation point.

The data show that at 12 months children's repertoire consisted of approximately the same small number of DG, RG and rw (slightly more numerous), while dw appeared at 15 months (in one child, in the other five children between 16 and 18 months). RG were present in the children's repertoire throughout all observations, but their mean number remained on average low, with minor oscillations across observation points (M= between 3,5 and 7,5). In contrast, the mean number of rw increased steadily, going from a value of 6.3 at 12 months to a value of 92 at 24 months.

As it could be expected, there was considerable individual variation, especially with respect to the repertoire of rw: three children (GIA, MAR, NIC) exhibited a more precocious pattern of vocal development, three children (ALE, GAL, SAV) a less precocious one. For example, NIC had a repertoire of 13 and 183 rw at, respectively, 12 and 24 months. At the same age points ALE, a child who developed at a slower pace, had 2 and 21 rw in his repertoire.

The total number of distinct RG identified across children and observation sessions was relatively high (N= 88), but three children (NIC, MAR and ALE) had fewer RG in their total repertoire (N= 5, 6 and 10, respectively), three children (GIA, SAV and GAL) markedly more (N= 22, 40 and 54 respectively). The range of variation for RG types was more constrained than for rw: the number of RG in each child's repertoire varied, depending upon the sample, between a minimum of 0–1 (in the three children with a smaller inventory of RG) and a maximum of 15–20 (in the three children with a larger RG repertoire). Very few, highly conventional RG were common to 4 or 5 children (e.g. NO, YES, BYE-BYE, BRAVO, SILENCE, GOOD), whereas the largest majority were idiosyncratic iconic RG specific of each child. The three DG we distinguished were present in the repertoire of all children at all observations, except in ALE's sessions at 23 and 24 months, in which only POINTING and REQUEST were represented.

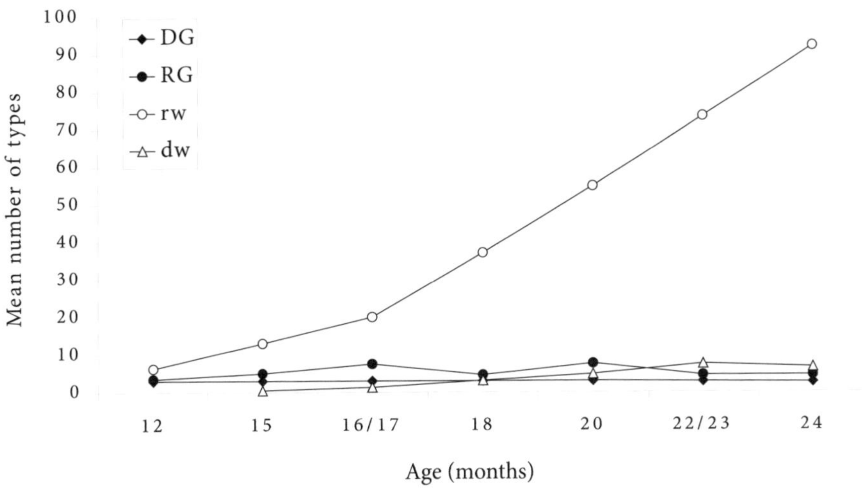

Figure 1. Children's repertoire

The individual differences noted in this as in other aspects of the children's production (see below) are similar to those reported in other studies of early gestural-vocal development (e.g. Capirci et al., 1996; Guidetti, 2002; Iverson et al., 1994; Volterra et al., 2005; see also Capirci, Contaldo, Caselli, & Volterra, 2005), and suggest the need of integrating analyses based on single cases with analyses based on larger groups.

One-element utterances

Figure 2 shows the development (mean number) of one-element utterances consisting of 1rw, 1RG, 1DG and 1dw.

The data show that on average, with the exception of the first session, the most frequent type of one-element utterances were always 1rw, which became very productive by 24 months (M=217). In contrast, 1RG utterances remained in fairly small number across all observation sessions (M comprised between 4 and 15). Limitations in use of 1RG utterances (compared to 1rw) were noted in all children, including those who had a fairly large gestural repertoire (GIA, SAV, GAL). In two children (NIC, GIA) 1rw utterances prevailed over 1RG and 1DG already at 12 months, in the other four children this occurred at 15 months. In agreement with findings reported by previous studies in which 1rw and 1RG utterances have been distinguished (e.g. Pizzuto, 2002; Pizzuto et al., 2005; Volterra et al., 2005), these data indicate that, in one element utterances, children's representational skills are expressed primarily and preferably through the vocal, not the gestural channel.

At 12 months 1DG utterances were almost as frequent (M = 22) as 1rw utterances (M = 28) but their mean number slowly decreased over time (M = 5 at 24

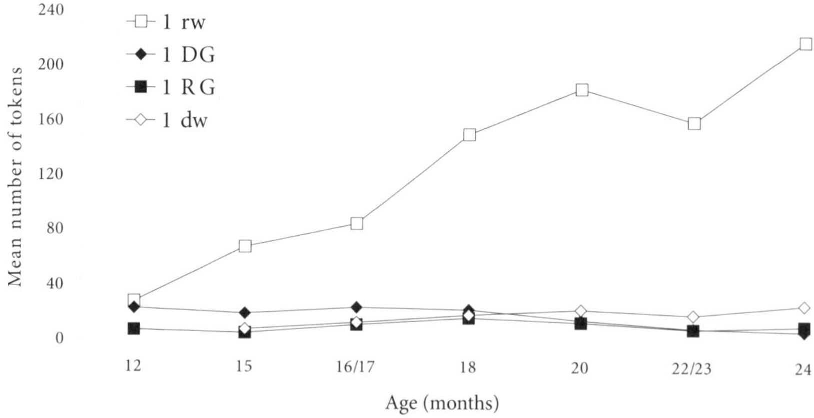

Figure 2. Children's one-element utterances

months) while, from 15 months on, utterances made of deictic words (1dw) were noted and, on average, slowly increased in time. However, 1dw utterances were produced in appreciable number only by two children — NIC and GIA — while they were less numerous in the other children.

Two-element utterances

Figure 3 provides an overview of the major types of two-element utterances observed in the children's production. In this figure the complementary and supplementary subtypes are grouped together, distinguishing the crossmodal ones (DG &,+ rw/dw) from the vocal only ones (rw &,+ rw/dw). Two-element combinations that occurred in very small numbers are not represented in the figure, but mentioned below.

It can be seen that in the earliest sessions the most frequent type of two-element utterances were bimodal equivalent combinations of a deictic gesture with a deictic vocalization (DG = dv, M at 12 and 16/17, respectively = 28 and 47). However, these subsequently declined and were in a very low number at 24 months (M = 6).

A different developmental pattern characterized crossmodal utterances of the complementary and supplementary types (DG &,+ rw/dw). These were present from 12 months, preceded in development vocal-only utterances of comparable information (rw &,+ rw/dw), increased steadily through 18 months, then decreased moderately and stabilized through the last three observation points, in coincidence with a marked increase of vocal-only utterances.

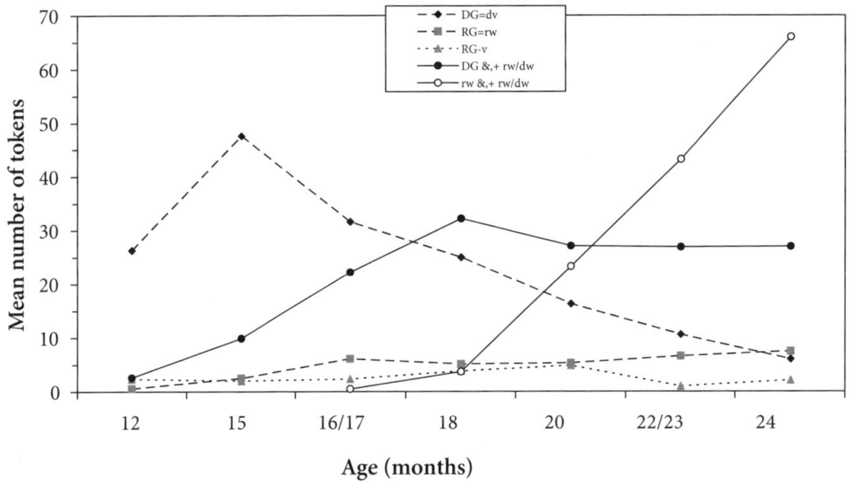

Figure 3. Children's two-element utterances

There was individual variation. For example, the increase of vocal-only utterances was more evident in three children (MAR and GIA and especially NIC), but less evident in the remaining three. The proportion of crossmodal vs. vocal-only utterances varied from child to child and, in each child, with development. However, these crossmodal combinations constituted a very salient feature of all children even after the appearance of vocal-only utterances: in five children these combinations were still the most frequent, or as frequent as vocal type at 22/23 months; in one child (NIC) vocal-only utterances prevailed since 20 months, but their crossmodal counterpart continued to be produced and even increased at 24 months.

Bimodal equivalent combinations of two representational elements (RG = rw) were produced from the earliest age points, increased slightly at 16/17 months, but their mean number remained low (between 2,5 and 7,6). Equally low was the mean number of representational gestures accompanied by a vocalization (RG-v). Finally, it must be noted that, similarly to what was found in other studies (e.g. Capirci et al., 1996; Pizzuto et al., 2005), two-element utterances consisting only of gestures were virtually absent (N = 3 across children and observations), and did not include any combination of two RG.

Comparing the gestural and vocal elements that appear in different types of utterances, relevant asymmetries can be noted in the distribution of deictic vs. representational elements in the gestural vs. the vocal modality. DG were used very productively not only in relatively simpler, bimodal combinations that expressed a generic deictic information (DG = dv), and declined with development, but also in more complex complementary and supplementary combinations (DG &,+ rw/dw) that increased and then stabilized in the children's production. Similarly to Capirci et al.'s (1996) study, POINT was by far the most frequent deictic gesture employed in these combinations, while SHOW and REQUEST were much less frequent.

In contrast, the use of RG was markedly more constrained: these appeared in bimodal equivalent, redundant combinations with rw (RG = rw), or in combinations with vocalizations that did not articulate their meaning (RG-v). The mean number of these combinations remained low across observations. Crossmodal supplementary combinations including representational gestures and words (e.g.. ALL_GONE + coco' <=chicken>) were noted, but were so infrequent (N = 28 tokens across children and observations) that their development could not be charted.

Quite differently, rw were productively used from the start in complementary and supplementary combinations with *either* another rw, *or* a dw. Since RG were attested in the children's repertoires (and in three children even in a sizeable number, as noted above), their extremely sparse use in supplementary crossmodal combinations, along with their limited use in one-element utterance, and the absence of utterances consisting of two RG, indicates that at these early stages of development children's use of gestures for representational purposes is considerably limited, and certainly differs from that observed in adults.

Complementary versus supplementary utterances

Figures 4 and 5 illustrate the distribution of crossmodal vs. vocal only combinations in the two major classes of complementary and supplementary utterances.

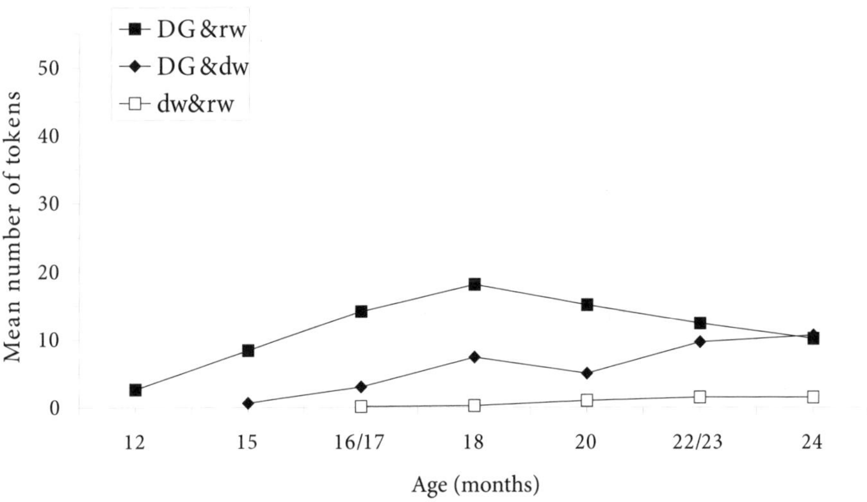

Figure 4. Children's complementary utterances

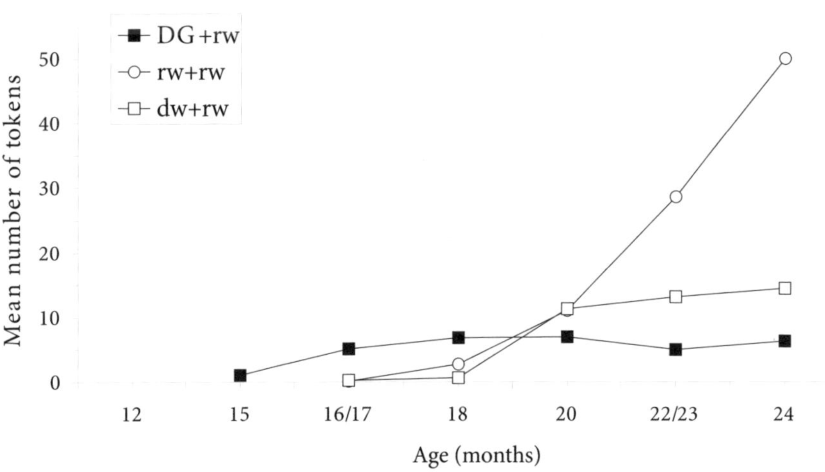

Figure 5. Children's supplementary utterances

The data indicate different developmental patterns for complementary vs. supplementary utterances. Within complementary utterances (Fig. 4), crossmodal combinations of a deictic gesture and a representational word (DG&rw) appeared earlier, increased and then moderately decreased over time, but remained markedly more frequent (M range = 2,6–18) than their vocal-only counterpart (dw&rw, M range: 0,1–1,5) through all observations. Vocal-only dw&rw combinations were in fact produced in appreciable number only by the child who exhibited the more advanced pattern of vocal development (NIC, from 16 months), they appeared later (20–24 months) and in small number in three other children (GIA, MAR, SAV), and were not produced by the two remaining children. Insofar as DG&rw utterances can be compared to a form of naming, these data suggest that, in the developmental period examined, naming is carried out for the most across modalities, rather than via two vocal elements (i.e. via dw&rw utterances).

Combinations of a deictic gesture and a deictic word (DG&dw), the type of deictic crossmodal utterance that is perhaps more common in adult usage, were present from 15 months on, and increased with development, reaching at 24 months the same mean number as DG&rw (M = 10). Individual developmental profiles again differed: these combinations were used at 15–16 months by two children (NIC and ALE), but appeared later (17–20 months) in the remaining four children, and were used with some frequency by only two children (NIC and GIA).

Within the set of supplementary utterances (Fig. 5), crossmodal combinations of a deictic gesture and a representational word (DG+rw) again preceded in development their vocal counterpart (dw+rw), increased moderately through 18 months, then decreased but somewhat stabilized in the last three samples. At the same time, there was a clear increase of dw+rw, and a more marked increase of utterances composed of two representational elements (rw+rw). Insofar as DG+rw utterances can be assimilated to a form of predication, these data suggest that predication is carried out initially across modalities but it is subsequently expressed preferentially in the vocal modality, more frequently via combinations of two representational words. Unlike what noted within complementary utterances, in supplementary utterances DG appear to be replaced, as development proceeds, by dw or rw.

Comparing Figures 4 and 5 it can also be noted that supplementary crossmodal combinations DG+rw appear after, and are less frequent than complementary combinations DG&rw. This data can be interpreted as an index of the greater complexity of the first compared to the second type of combinations, in agreement with proposals formulated in other studies (e.g. Capirci et al., 1996; Goldin-Meadow & Butcher, 2003; Pizzuto et al., 2005).

At a global level, across children and observation sessions, there were relevant asymmetries in the distribution of crossmodal vs. vocal utterances in the complementary or supplementary classes. *Crossmodal* utterances fell primarily in the complementary class (69%), and were markedly less frequent in the supplementary class (31%). The reverse pattern held for *vocal-only* utterances: the vast majority (96%) were supplementary, a very small proportion (4%) complementary. These distributional patterns are very similar to those reported, for a smaller set of longitudinal data, by Pizzuto et al. (2005). They suggest that children's crossmodal productions are somewhat biased towards conveying complementary information, while their vocal only utterances are biased towards conveying supplementary information.

Summary and concluding remarks

The data we have described corroborate and extend the findings from previous work (Capirci et al., 1996; Pizzuto, 2002; Volterra et al., 2005; Pizzuto et al., 2005), and provide new information on the link and differences between deictic and content-loaded elements in children's developing gesture-speech system during the second year of life.

Recalling the questions we initially formulated, our data evidence an uneven distribution of deictic vs. representational elements in the gestural as compared to the vocal modality. This is observed in both the children's repertoires and, more significantly, in their utterance patterns.

With respect to the repertoires we found that on average, in the children we examined, vocal representational elements (rw) markedly prevailed over gestural ones (RG) from 15 months on. RG were present in all children's repertoires through all observation sessions, with individual variation similar to that reported by other studies of children's early gestures (Capirci et al., 1996, 2005; Guidetti, 2002). However their mean number remained low while at the same time, and quite obviously for children acquiring a spoken language, rw steadily increased. Thus, in agreement with what was observed by Volterra et al. (2005), the "shift to the vocal modality" that is observed in children's repertoires during the second year of life is not due to a contraction of children's gestural repertoire but to a parallel, and much greater expansion of their vocal repertoire (see also Capirci et al., 2005).

DG preceded in development dw, and were present in all children's repertoires across observation sessions, with no significant individual variation. This was especially true for the POINT gesture.

The analysis of utterance patterns confirmed and clarified the different distribution of deictic vs. representational elements in children's production. The strong links between gestural and vocal behaviors were evident from the start in the frequent use of vocalizations with gestures (especially with DG) and, more significantly, in gesture-word combinations of the complementary and supplementary types.

DG, like dw and unlike RG and rw, constitute a limited set. In principle, if a productive use of gestures was linked primarily to the number of gestures in children's repertoires, one could expect that RG would be used with at least the same frequency as DG, especially by children who have a fairly rich repertoire of RG (as was the case, in our sample, of GAL, GIA, SAV). But in this study as in previous work we found that this is not the case. Regardless of the number of RG they possessed, the children we examined did *not* use them productively in their one- and two-element utterances. One-element utterances were realized primarily by means of rw. In two-element utterances RG occurred almost exclusively in redundant, bimodal combinations with rw. As noted earlier, these productions can be assimilated to one-element utterances in which a gesture and a word are superimposed, rather than combined, sorting the effect of 'reinforcing' or 'magnifying' one and the same piece of information (e.g. NO=no). Additional research is certainly needed to clarify the functions (and developmental course) of these productions.

The gestures children employed very productively in true, complementary and supplementary combinations with words, notably with rw, were DG, a pattern which is not observed in adult communication — and which, as remarked in Pizzuto et al. (2005), is obscured if deictic elements are not distinguished from representational ones. The robustness of this feature in children's early gesture-speech system is highlighted by the fact that these combinations were productive even in children who had a fairly rich repertoire of both rw and dw, as was the case for NIC, MAR, GIA.

Crossmodal complementary and supplementary combinations of DG and rw can be compared, as suggested here and elsewhere, to vocal-only combinations of a dw and a rw that allow children to express two basic functions of human language: "naming" and "predicating". These two functions of human language appear to develop first (and also more productively in the case of "naming") across modalities, and only later within the vocal modality only. However, there was a clear prevalence of vocal representational elements in two-element utterances expressing supplementary information.

Taken together, these patterns underscore, on one hand, the relevance of gestural deixis as a primary (albeit ancillary) device in early language development, on the other hand the limitations of children's representational abilities in the gestural modality. In these early stages speech and gesture are linked primarily via the use of deictic gestures. These undoubtedly provide relevant information but do not "articulate" the information encoded in children's early crossmodal combinations in the same manner in which iconic and metaphoric gestures do in adult gesture-speech system. In other words, a mature use of gestures for representational purposes does not appear to be an early achievement.

More extensive longitudinal and cross-sectional studies are certainly needed to explore the development of more mature uses of representational (and also beat) gestures as children get older, and more accurate distinctions of the typology and functions of content-loaded gestures need to be made (see for example Guidetti, 2002). The broad typological differences we explored and documented are just one step towards a more comprehensive understanding of children's developing gesture-speech system in the early stages.

Notes

1. Throughout this paper labels for gestures are given in capital letters.

Acknowledgements

We gratefully acknowledge partial financial support from the European Science Foundation EUROCORES Programme OMLL, funds from the Italian National Research Council and the EC Sixth Framework Programme under Contract no. ERAS-CT–2003–980409. We also wish to thank Silvia Baldi, Ausilia Elia, Oriana La Veneziana and Elisa Nesti for their substantial help with data collection, transcription and coding.

References

Bates, Elizabeth (1976). *Language and context. The acquisition of pragmatics*. New York: Academic Press.

Bates, Elizabeth, Laura Benigni, Inge Bretherton, Luigia Camaioni, & Virginia Volterra (1979). *The emergence of symbols: Cognition and communication in infancy*. New York: Academic Press.

Bates, Elizabeth, Luigia Camaioni, & Virginia Volterra (1975). The acquisition of performatives prior to speech. *Merril Palmer Quarterly, 21* (3), 205–226.

Blake, Joanna (2000). *Routes to child language — Evolutionary and developmental precursors*. Cambridge: Cambridge University Press.

Brown, Roger (1973). *A first language*. Cambridge, MA: Harvard University Press.

Butcher, Cynthia & Susan Goldin-Meadow (2000). Gesture and the transition from one- to two-word speech: When hand and mouth come together. In David McNeill (Ed.), *Language and gesture* (pp. 235–257). Cambridge: Cambridge University Press.

Butterworth, George (2003). Pointing is the royal road to language for babies. In Sotaro Kita (Ed.), *Pointing: Where language, culture and cognition meet* (pp. 9–33). Mahwah, NJ: Lawrence Erlbaum.

Capirci, Olga, Jana M. Iverson, Elena Pizzuto, & Virginia Volterra (1996). Gestures and words during the transition to two-word speech. *Journal of Child Language, 23*, 645–673.

Capirci, Olga, Annarita Contaldo, M. Cristina Caselli, & Virginia Volterra (2005). From action to language through gesture: A longitudinal perspective. *Gesture*, 5 (1/2), 155–175. (This volume)

Capobianco, Micaela (2006). Gesti, parole e prime combinazioni nello sviluppo tipico e primo confronto con bambini nati pretermine. Doctoral Dissertation, University of Rome "La Sapienza".

Capone, Nina C. & Karla K. McGregor (2004). Gesture development: A review for clinical and research practices. *Journal of Speech, Language and Hearing Research*, 47, 173–186.

Clark, Eve V. (2003). *First language acquisition*. Cambridge: Cambridge University Press.

Deacon, Terrence (1997). *The symbolic species — The coevolution of language and the human brain*. London: Penguin Press.

Goldin-Meadow, Susan & Cynthia Butcher (2003). Pointing toward two-word speech. In Sotaro Kita (Ed.), *Pointing: Where language, culture and cognition meet* (pp. 85–107). Mahwah, NJ: Lawrence Erlbaum.

Goldin-Meadow, Susan & Marolyin Morford (1985). Gesture in early child language: Studies of hearing and deaf children. *Merrill-Palmer Quarterly, 31*, 145–176.

Goldin-Meadow, Susan & Marolyin Morford (1990). Gesture in early child language. In Virginia Volterra & Carol J. Erting (Eds.), *From gesture to language in hearing and deaf children* (pp. 249–262). Berlin: Springer-Verlag. (1994 — 2nd Edition Washington, D.C.: Gallaudet University Press).

Greenfield, Patricia M. & Smith, Joshua H. (1976). *The structure of communication in early language development*. New York: Academic Press.

Guidetti, Michèle (2002). The emergence of pragmatics: Forms and functions of conventional gestures in young French children. *First Language*, 22, 265–285.

Iverson, Jana M., Olga Capirci, & M. Cristina Caselli (1994). From communication to language in two modalities. *Cognitive Development*, 9, 23–43.

Kendon, Adam (1996). An agenda for gesture studies. *Semiotic Review of Books*, 7 (3), 8–12.

Kita, Sotaro (Ed.) (2003). *Pointing: Where language, culture and cognition meet*. Mahwah, NJ: Lawrence Erlbaum.

Lock, Andrew J. (1980). *The guided reinvention of language*. London: Academic Press.

Lyons, John (1977). *Semantics (vols. 1 &2)*. Cambridge: Cambridge University Press.

Mayberry, Rachel I. & Elena Nicoladis (2000). Gesture reflects language development: Evidence from bilingual children. *Current Directions in Psychological Science*, 9 (6), 192–196.

McNeill, David (1992). *Hand and mind — What gestures reveal about thought*. Chicago: University of Chicago Press.

McNeill, David (Ed.) (2000). *Language and gesture*. Cambridge: Cambridge University Press.

Newport, Elissa & Richard P. Meier (1985). The acquisition of American Sign Language. In Dan I. Slobin (Ed.), *The crosslinguistic study of language acquisition, Vol. 1: The data* (pp. 881–938). Hillsdale, NJ: Lawrence Erlbaum.

Pizzuto, Elena (2002). Communicative gestures and linguistic signs in the first two years of life. Paper presented at the *EURESCO Conferences, Brain Development and Cognition in Human Infants*, Maratea, Italy, June 7–12 2002.

Pizzuto, Elena, Micaela Capobianco, & Antonella Devescovi (2005). Gestural-vocal deixis and representational skills in early language development. *Interaction Studies, 6*, 223–252.

Slobin, Dan I. (Ed.) (1985). *The crosslinguistic study of language acquisition. Vol. 1: The data, Vol. 2: Theoretical issues.* Mahwah, NJ: Lawrence Erlbaum.

Volterra, Virginia, M. Cristina Caselli, Olga Capirci, & Elena Pizzuto (2005). Gesture and the emergence and development of language. In Michael Tomasello & Dan I. Slobin, (Eds.), *Beyond nature-nurture — Essays in honor of Elizabeth Bates* (pp. 3–40). Mahwah, N.J.: Lawrence Erlbaum.

Volterra, Virginia & Carol J. Erting (Eds.) (1990). *From gesture to language in hearing and deaf children.* Berlin: Springer Verlag. (1994 — 2nd Edition Washington, D.C.: Gallaudet University Press).

Volterra, Virginia & Jana M. Iverson (1995). When do modality factors affect the course of language acquisition? In Karen Emmorey & Judy Reilly (Eds.), *Language, gesture, and space* (pp. 371–390). Hillsdale, NJ: Erlbaum.

A cross-cultural comparison of communicative gestures in human infants during the transition to language*

Joanna Blake, Grace Vitale, Patricia Osborne, and Esther Olshansky
York University

The entire bodily gestural repertoire of four different infant groups was coded over the age period of 9 to 15 months. Two small samples of English-Canadian and Parisian-French infants were filmed every two weeks at home. A larger sample of Japanese infants was visited for 7 sessions and of Italian-Canadian infants for 4 sessions at 9 months and 15 months and again at 3 years. Language measures were collected for the last two groups. Increases in Comment gestures, particularly pointing, in Object exchange gestures, and in Agency gestures were found across almost all groups. Decreases in Reach-request and in Emotive gestures were also found for most groups. The increasing group of gestures was positively related to vocabulary acquisition, particularly to receptive vocabulary. Reach-request and Protest gestures at 15 months were negative related to different aspects of language at 3 years. The importance of examining the entire nonverbal communicative repertoire across cultures is discussed in terms of assessing the relationship of gestures to language acquisition. Changes in the gestural repertoire appear to be universal across infants of different cultures, at least those examined.

Nonverbal gestural communication, especially pointing, has been related to the development of verbal communication (e.g., Bates, Benigni, Bretherton, Camaioni, & Volterra, 1979; Camaioni, Caselli, Longobardi, & Volterra, 1991; Desrochers, Morisette, & Ricard, 1995; Rowe, 2000). However, few studies have charted the entire gestural repertoire over the transitional period to language in order to determine precise changes in particular gestures that might bear on language. Apart from our own work, there is only the study of two male infants by Zinober and Martlew (1985). There is also little information about the universality of the infant gestural repertoire, and of the changes with age in the repertoire, across different cultures. The research summarized in this paper was directed at these two issues.

Ekman and Friesen (1969) first made a distinction between informative and expressive or adaptive behaviors. Informative behaviors are typically limb movements "which seem designed to transmit information" (p. 52) and "have some shared decoded meaning… in a set of observers" (p. 55). Expressive or adaptive behaviors may involve less awareness and are often used to manage emotions. In the repertoire of gestures that we studied, we included both informative (termed here Comment) and expressive (called here Emotive) gestures. Previous research has tended to focus on informative gestures, particularly pointing which increases in the second year (Blake, 2000; Guidetti, 2002; Iverson, Capirci, & Caselli, 1994; Lempers, 1979; Leung & Rheingold, 1981; Murphy & Messer, 1977; Zinober & Martlew, 1985). It is generally agreed that pointing has two functions: request and declarative. Its request function makes it interchangeable with reaching, but the definition of pointing typically requires an extended index finger, usually with other fingers curled and arm extended (see Blake, 2000; Butterworth & Morisette, 1996). However, there are cultures that point with the whole hand extended, as in reaching (Wilkins, 2003). The declarative function of pointing has as its motive the sharing of information with the observer, i.e., the informative function of Ekman and Friesen (1969). It seems clear that in human infants, pointing by 12 months has a declarative (informative) motive and that infants understand that pointing can direct another person's attentional state (Legerstee & Barillas, 2003; Liszkowski, Carpenter, Henning, Striano, & Tomasello, 2004). They pointed more frequently to an adult who did not look at an event or object of interest. They thus wanted to share interesting events with people, but they did not do this with objects (e.g., a doll) (Legerstee & Barillas, 2003). Declarative pointing has also been linked to understanding of another's intentions (Camaioni, Perucchini, Bellagamba, & Colonnesi, 2004). Children who produced more declarative points early in the second year completed more of the experimenter's incompleted actions in a Meltzoff-type task (Meltzoff, 1995).

Pointing is a good predictor of word acquisition (Bates, et al., 1979; Camaioni et al., 1991; Rowe, 2000). The age of onset of communicative pointing has been related to both expressive and receptive language at 24 months (Desrochers et al., 1995). Spanish toddlers continued to use pointing with a deictic word even though they were able to use the word alone, so that the redundancy may have helped them to localize the target object (Rodrigo, González, de Vega, Muñetón-Ayala, & Rodriguez, 2004).

More general measures of gestures have also been related to language acquisition. A gestural complex, composed of pointing, showing, giving and ritual request (so-called deictic gestures), as reported by parents, was more strongly related to language comprehension than to language production (Bates, Thal, Whitesell,

Fenson, & Oakes, 1989). The sum of actions and gestures reported by parents on the Finnish version of the MacArthur Communicative Development Inventory (MCDI), an inventory which asks parents which actions and gestures their infant had produced, predicted receptive scores on the Reynell Developmental Language Scale at 18 months and expressive language scores on the Bayley Scale at 24 months (Laakso, Poikkeus, Katajamaki, & Lyytinenen, 1999). These scales are administered to infants by an experimenter and involve either actions performed by the infant in accordance with instructions (comprehension) or elicitation of verbal production. The sum of gestures, alone, on the MCDI was correlated with comprehension on the Japanese version of the MCDI, with age and production partialled out (Ogura & Watamaki, 2001), although in this case, both scores were provided by parents.

In a longitudinal study across the same age period (9 to 15 months) as the studies reviewed below, Carpenter, Nagell, and Tomasello (1998) found that show emerged first, followed by give, and then point, and declarative gestures before imperative gestures for most infants. However, while the functions were discriminated (declarative vs. imperative), reaches were subsumed under points. The age of emergence of proximal declarative gestures (give, show) was related to that of flexible, referential words.

Gestures also are important in the transition to sign language in both deaf children and hearing children exposed to sign language by their parents. A single hearing child of deaf parents produced more types of communicative gestures and more symbolic gestures than deictic gestures (show, point, request) by comparison with a group of monolingual (non-signing) hearing children (Capirci, Iverson, Montanari, & Volterra, 2002). By 20 months, he produced many more gesture plus word combinations than any of the comparison children.

Continuity between prelinguistic gestures and linguistic signs can be studied by determining if the most frequent values in prelinguistic gestures are also the most frequent in early signs. In a study of 4 deaf children of deaf parents, the prelinguistic gestures coded at 7–8, 10, and 13 months were reach, point, show, gimme, wave, conventional (e.g., patty cake) and symbolic gestures. These gestures were similar to early signs (i.e., continuous) in handshape, in palm orientation, in hand-internal movement, and in hand arrangement. They were dissimilar (i.e., discontinuous) in place of articulation and in path movement (Cheek, Cormier, Repp, & Meier, 2001).

The research reviewed reveals that nonverbal communicative behaviors provide a foundation for verbal communicative behaviors in both words and signs. It is clear, however, that nonverbal communication, even when restricted to body movements, has been quite narrowly defined by most investigators. The repertoire of gestures used by human infants is quite large (see also Zinober & Martlew, 1985). The question then arises as to which of these are positively related to language acquisition and which are negatively related? Furthermore, the studies reviewed have tended to focus on one culture. Thus, although in some cases where the coding systems are similar, we can compare across cultures, this has never been done for the entire repertoire of communicative body movements. The purpose of the studies summarized below was to begin to provide cross-cultural information on the development of infants' gestural repertoire. The cultures compared were English-Canadian, Italian-Canadian, Parisian-French, and Japanese. The major question of this research, then, is whether or not we can make a case for the universality of the early gestural repertoire or whether universality applies to only restricted aspects of the repertoires. It might also be expected that the three Western cultures included might be more similar to each other than to the Japanese sample.

Secondly, increases and decreases in the repertoire with development were also of interest. It was predicted that gestures involving the sharing of objects (Object exchange) and the sharing of information (Comment), as well as specific Request gestures showing cognizance of agency, would increase across all cultures (see Blake, 2000). These gestures were expected to increase over the transition period to language because they are important for language acquisition. They were predicted to relate positively to measures of language. Primitive gestures that potentially interfere with verbal communication, such as Reach-request, Protest, and Emotive gestures, were expected to decrease in all cultures over the transition period to language. They were predicted to relate negatively to measures of language.

Method

Participants

There were two small samples of infants observed intensively, i.e. every two weeks, at home: 5 English-Canadian infants, 3 female and 2 male, and 4 Parisian-French infants, 2 female and 2 male. These samples were observed from 9 to 14 months, with the exception of 1 male from each sample who was observed from 11 months to 14 months.

There were two larger samples of infants also observed at home. Twelve Japanese infants, 6 male and 6 female, living in the greater Tokyo-Yokohama area entered the study at different ages from 9 months (4 infants), 10 months (5 infants), or 11 months (3 infants) and completed it between 12 and 14.5 months, for a total of 7 sessions. Thirty Italian-Canadian infants, 15 males and 15 females, were visited four times: at 9 months, at 9.5 months, at 15 months, and at 37 months. The first three sessions were unstructured observations.

All infants were full-term and living in two-parent families (with one exception). In the small samples and the Japanese sample, all but one were first-born. Seventeen of the Italian-Canadians wer first-born. The parents of infants in the small samples were highly educated. In the larger samples, level of education varied but the families were middle class.

Procedure

Infants were filmed at home in naturalistic interaction with a parent, usually the mother or a caretaker (2 French infants). The sessions ranged from 20 to 60 minutes (the longest only for 1 infant), but were typically 30 minutes long. The only toys provided by the observer were a book and a wind-up toy in an attempt to elicit point-in-book and give-request gestures (see Table 1).

Measures

Measures of language were administered to the larger samples. Japanese mothers completed the Japanese version of the MacArthur Communicative Development Inventory (JCDI) (Ogura & Watamaki, 1998) at a session close to 12 months. Only the total numbers of words understood and produced were used, including words in baby talk form because Japanese children produce many of these (Ogura & Watamaki, 2001).

Italian-Canadian mothers provided a list of words that their infants had produced by 15 months, to which were added any words heard on the videorecording. Words with the same meaning in Italian and English were counted once. At 37 months, these children were given measures of communicative competence, of expressive language, and of receptive language. The measure of communicative competence was a board game based on "Old McDonald's Farm," that was unfamiliar to them. The game involved a sequence of spinning a wheel, rolling a die with pictures on it, matching the picture of the animal face-up on the die to a puzzle piece, and placing that puzzle piece in a puzzle board. The children were taught the game and then asked to explain it to a puppet. Their explanations were scored on a scale according to the level of accuracy and specificity with which the four main steps were described. (See Blake, Olshansky, Vitale, & Macdonald, 1997, for more detail.) Reliability of the scoring on 30% of the sample was $r = .97$, $p = .0001$. The measure of expressive language was a spontaneous 30-minute speech sample recorded, transcribed, and scored for mean length of utterance (MLU) following

Table 1. Inventory of Communicative Gestures

Comment	
Point:	Involves the extended arm and index finger, with lightly or tightly curled other fingers. No contact with object. Directed at person, object, or event.
Point-in-book:	Same as above except that the point is in a book so can be contact.
Wave bye-bye:	Waving of arm or hand.
Head shake/nod:	Can be yes or no, but latter must be with positive affect; otherwise belongs in protest.
Show:	Holds **up** object to another, usually with elbow bent. There is movement of infant's arms, but object remains in her possession.
Object exchange	
Give:	Infant hands object to another. There is movement of arms, and object changes hands. Not a request.
Take:	Infant reaches and obtains an object from another. There is movement of the infant's arms, and object changes hands. Also coded when the infant accepts an object given by another.
Offer:	Infant holds **out** an object to another, usually with arm extended and palm up. There is movement of the infant's arm, but object remains in her possession. Not a request.
Request	
Reach:	Arm is extended, palm is usually down, hand is open, and fingers are straight. Usually continued until request is granted. This reach is for something out of reach. Reaching that culminates in the infant's grasping something on his/her own is not coded.
Point:	Same as above, but context is that of request.
Up:	One or both arms are raised towards another, or merely moved away from the body to allow room for another's hands to pick up infant (in response to adult's motion towards infant with hands out). Includes reach to adult if results in being picked up.
Down:	Infant leans away from another and towards ground if held in arms. If in a lap or seat, may lean and/or stiffen and slide down.
Agency gestures	
Seek assistance:	Infant exhibits effortful action, and affect is often negative. Must be eye contact.
Give/offer:	As before, but wants parent to do something with object.
Take hand:	Infant takes part of another's body, usually the hand, and either guides it to an object or leads the person somewhere to do something (e.g., open a door).
Protest/Rejection	
Turn away:	Infant turns head or body away from another's actual or approaching physical contact.
Push/pull away:	Infant uses hand or arm to push away object being offered or another person. Infant attempts to remove self, hand, foot, etc. from hold of another.
Resist bodily:	Includes kick, body stiffening, movement away, involving infant's whole body.
Hit:	Open hand is struck against object being offered or another person.
Head shake:	Movement of head from side to side. With negative affect.
Emotive	
Flap/wave arms:	Arms sometimes extended, sometimes bent at elbow, moved in an up-and-down or side-to-side direction. No object in hands.
Clasp/clap hands:	Open hands brought together.
Bounce:	Bodily movement up and down, while either sitting or standing.

rules outlined in Blake, Quartaro, and Onorati (1993). Reliability on MLU for 20% of the sample was $r = .99$, $p = .0001$. The measure of receptive language was the Peabody Picture Vocabulary Test (PPVT) (Dunn & Dunn, 1981) in both English and Italian. All children with one exception scored higher on the English form, so these scores were used except for this one child.

Coding System

The inventory used to code infant gestures included 5 general categories with more specific categories included in each. Gestures restricted to object manipulation that did not involve social interaction were excluded. Also excluded were symbolic or depictive gestures. Some conventional gestures (head nod/shake, waving 'bye'bye, clap hands) were included in this inventory, though some researchers consider them to be representational (e.g., Iverson & Thal, 1998). Conventional gestures are those that do not change form with development, and they have been investigated as a separate category (including pointing) in children between 16 and 36 months (Guidetti, 2002). Our inventory includes only movements of the body (head, limbs or hands) that communicate meaning to the observer (See Table 1). Reliability on this inventory ranged in the studies from 84% agreement on 18–30% of the sessions to Cohen's kappa = 0.89–0.94 on 13–33% of the sessions.

Results

For almost all groups, Comment, Object exchange, and Agency gestures did increase across the sessions (see Figs. 1, 3, and 4) as predicted. Most of the increase in Comment gestures was in declarative pointing (see Fig. 2); showing changed very little. For the Italian-Canadian group, the increases between the two ages, 9.5 and 15 months, for Comment and Object exchange gestures were significant, $F(1, 290) = 6.82$, $p<0.01$ and $F(1, 290) = 11.81$, $p<0.01$, respectively (Blake et al., 1997). The increase in Agency gestures was not significant. The English-Canadian group, alone, did not show a significant increase in Comment gestures because their relative frequency of these was already quite high at 10 months. They increased somewhat in Object exchange gestures but not significantly. Their increase in Agency gestures between 9.5 and 14 months was significant, however, $t(3) = 4.46$, $p<0.02$. For the French infants, Object exchange gestures across 4 age periods (9–10 mos., 10½–11½ mos., 12–13 mos., 13½–14½ mos.) showed both a linear trend, $F(1, 3) = 25.27$, $p=0.02$, and a cubic trend, $F(1, 3) = 14.17$, $p=0.03$ (Blake & Dolgoy, 1993), because these gestures increased dramatically at 12 months and then decreased

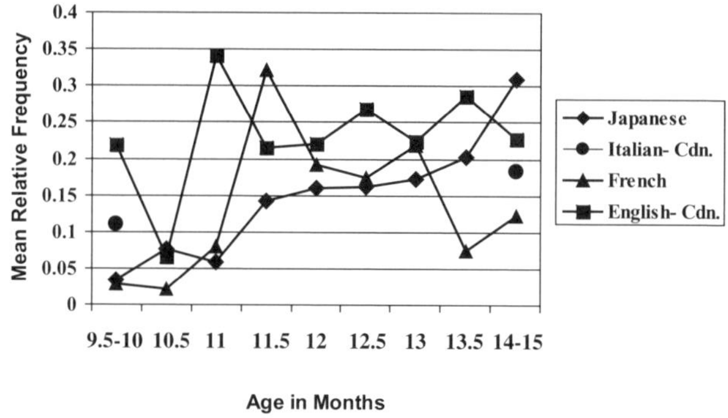

Figure 1. Comment Gestures

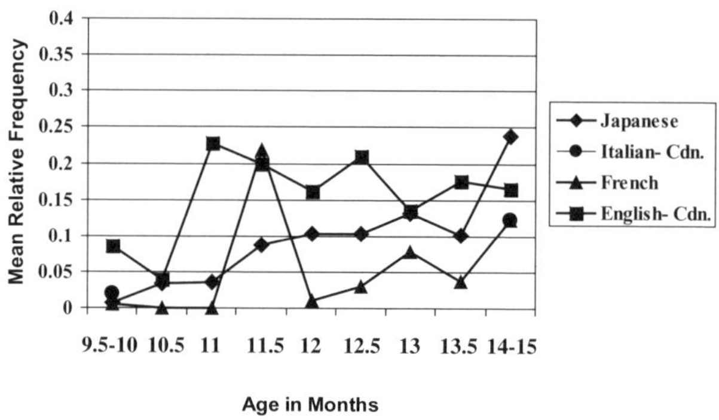

Figure 2. Declarative Pointing Gestures

after 13 months. Comment gestures showed the same curvilinear trend, but it was not significant. While the extreme points for Agency gestures (9.5 and 14 months) are quite different for French infants, the difference between them was not significant. For Japanese infants, Comment gestures increased linearly across sessions, $F(1, 10) = 30.70$, $p = 0.0001$, as did declarative pointing gestures, $F(1, 10) = 20.84$, $p = 0.001$ (Blake, Osborne, Cabral, & Gluck, 2003). Like the French infants, Object exchange gestures appear to show a curvilinear trend, but this was not significant. A significant linear trend was found for Agency gestures across sessions, $F(1, 10) = 6.38$, $p = 0.03$ (Blake et al., 2003). These linear trends may not be apparent for the Japanese infants on the Figures because they are plotted against Age rather than Session. The number of infants represented at each age varied from 5 to 12 for the Japanese sample.

In summary, the predicted increases were confirmed for the general category of Comment, including declarative pointing, for Italian-Canadian infants and for Japanese infants and for the specific category of declarative pointing for Japanese

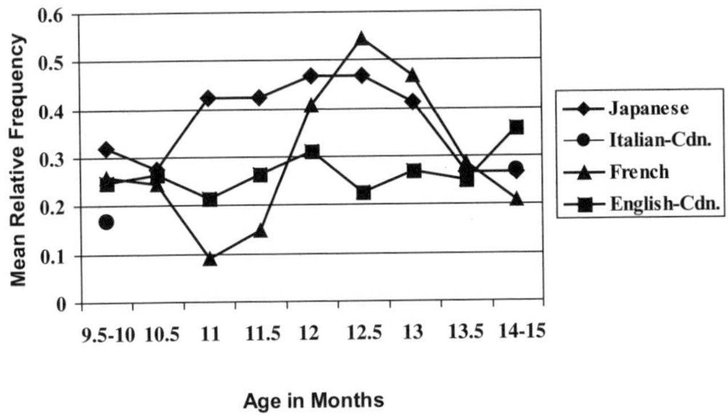

Figure 3. Object exchange Gestures

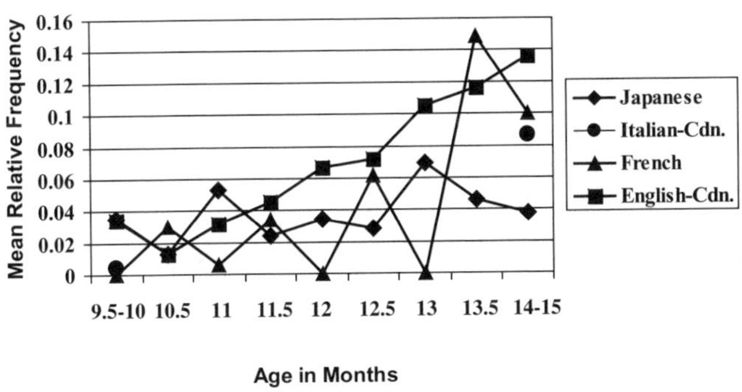

Figure 4. Agency Gestures

infants. For the French infants, Comment gestures, and declarative pointing, increased until 11.5 months and then decreased. For the English-Canadian infants, the predicted increases in Comment gestures, and in declarative pointing, were not confirmed. There were increases in both the general and specific categories at 11 months; but the infants already had a relatively high level of both at the earliest age and the increase was not significant. The predicted increase in Object exchange gestures was confirmed for Italian-Canadian infants and for French infants; while for Japanese infants, these gestures increased and then decreased after 13 months. English-Canadian infants showed an increase in these gestures, but it was not significant. The predicted increase in Agency gestures was confirmed for English-Canadian infants and for Japanese infants. Italian-Canadian and French infants increased in these gestures but not significantly.

In terms of the predicted declines in Reach-request, Protest, and Emotive gestures (see Figs. 5, 6, and 7), the Italian-Canadian infants decreased in Reach-request

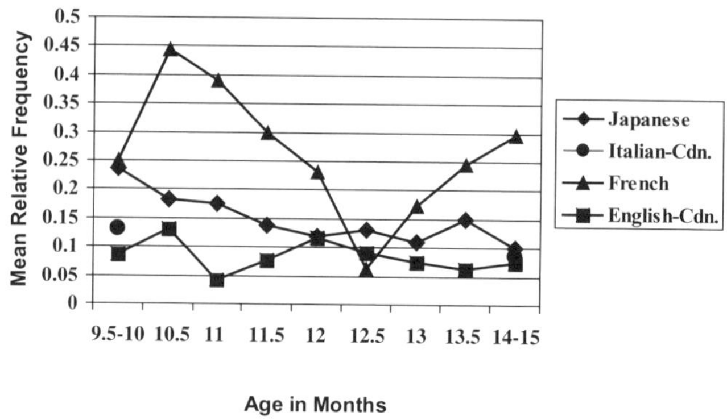

Figure 5. Reach-request Gestures

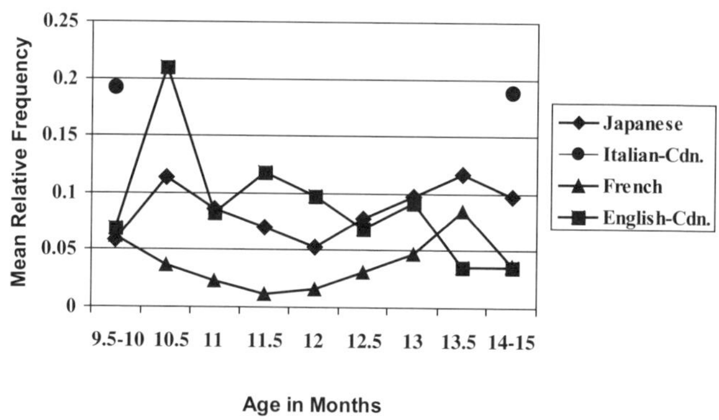

Figure 6. Protest Gestures

and Emotive gestures but not significantly; they did not decline at all in Protest gestures. English-Canadian infants declined only in Emotive gestures but then increased slightly after 12 months; there was an almost significant quadratic trend across ages, $F(1, 8) = 137.73$, $p = 0.054$. For French infants, there was a significant linear decrease in Emotive gestures across 4 age periods, $F(1, 3) = 20.47$, $p = 0.02$ (Blake & Dolgoy, 1993). For Japanese infants, Reach-request gestures showed a significant linear decrease across sessions, $F(1, 10) = 7.72$, $p = 0.02$ (Blake et al., 2003).

In summary, the predicted decline in Reach-request gestures was confirmed for Italian-Canadian infants and for Japanese infants, but not for French infants or for English-Canadian infants. French infants did show a decrease in these gestures and then an increase after 12.5 months. The predicted decline in Emotive gestures was confirmed for French infants and for English-Canadian infants, who increased somewhat in these gestures after 12 months. Italian-Canadian infants

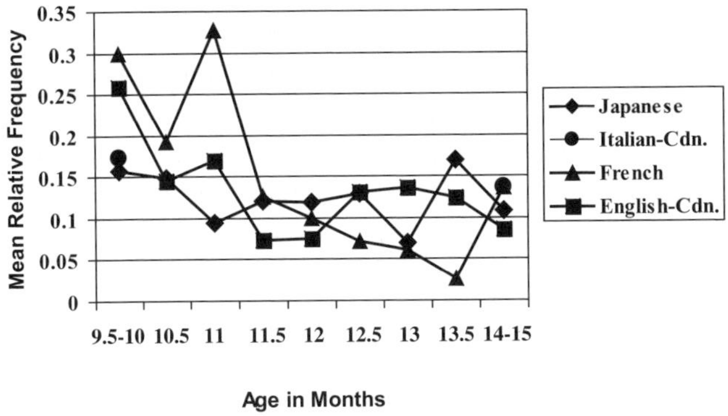

Figure 7. Emotive Gestures

showed a decrease in these gestures, but it was not significant. The predicted decline in Protest gestures was not confirmed for any group.

Relations between Gestures and Language measures

For the Italian-Canadian infants, the relative frequency of Object exchange at 15 months was positively correlated with productive vocabulary size at the same age, as reported by mothers ($r = .44, p < 0.05$). The relative frequency of Comment gestures at 15 months predicted receptive vocabulary on the PPVT at 37 months ($r = .37, p < 0.05$). The relative frequency of Reach-request gestures at 15 months was negatively related to MLU at 37 months ($r = -.42, p < 0.05$). The relative frequency of Protest gestures at 15 months was negatively related to communicative competence (game scores) at 37 months ($r = -.48, p < 0.01$) (see Blake et al., 1997). For Japanese infants, the relative frequency of Agency gestures after 12 months tended to be related to receptive vocabulary as reported by mothers on the JCDI ($r = .55, p = 0.06$) (Blake et al., 2003).

Discussion

In assessing universality of the gestural repertoire, it should be stressed that the English-Canadian and French samples were small and, therefore, not very representative. Also the Japanese infants entered the study at somewhat different ages. Finally, the Italian-Canadian sample was observed only at the first and last ages of the other samples. Nevertheless, as summarized in the Results, the increases and decreases in gestural categories were remarkably similar across cultures. Some differences in the relative frequencies of certain gestures were apparent, however.

English-Canadian infants were somewhat higher at most ages in their use of Comment and declarative pointing gestures compared to the other groups, except for the oldest age when Japanese infants surpassed them. They were also higher in their use of Agency gestures at most ages from 12 months. Japanese infants engaged in much give-and-take with their mothers and were higher in Object exchange than other groups at most ages, except for French infants after 12 months. Across these gesture categories, French infants appear to display the most irregular functions, perhaps reflecting variability within this small group.

English-Canadian infants were lowest in the use of Reach-request gestures at the earliest age and French infants highest except at the youngest age and at 12.5 months. French infants were also highest in their use of Emotive gestures at the first three ages. At the first and last ages (the only ages at which they were observed), Italian-Canadian infants were highest in Protest gestures. The differences between groups at specific ages were not tested because of the different-sized samples and because the major interest was in whether or not the patterns of change were similar.

Despite the belief that Italians are highly gestural (and more than half of the Italian-Canadian mothers spoke mostly Italian to their infants), Italian-Canadian infants were quite similar to the other groups in their use of gesture, except Protest. They were *not* more similar than other groups to English-Canadian infants, who share their larger culture. Japanese infants, a non-Western culture, were also quite similar to other groups, except in their use of Object exchange gestures. We would thus hypothesize that our infant gestural repertoire is a universal one and that any differences between groups, particularly in the use of declarative pointing and give-and-take, are likely attributable to differences in parental input. We are currently examining maternal gestures to these infants. While such parental differences would reflect cultural differences, the overall picture appears to be that these are minor influences in the infant gestural repertoire. They are more important, of course, in the acquisition of conventional gestures. There are three that are a part of our repertoire: head nod/shake, 'bye 'bye, and clap hands. All groups exhibited these gestures, but they are late-appearing (Blake, 2000).

The first set of gestures, Comment, including declarative pointing, Object exchange, and Agency, was predicted to relate positively to measures of language. The second set of gestures, Reach-request, Protest, and Emotive, was predicted to relate negatively to measures of language. These predictions could be tested only for the larger samples, Japanese and Italian-Canadian. For the Japanese infants, only Agency gestures after 12 months tended to be significantly related to receptive vocabulary size as reported by mothers. For the Italian-Canadian infants, Object exchange gestures were concurrently related to productive vocabulary size as reported by mothers. More interesting for this group were the significant predictions of language measures at 3 years. Comment gestures (as a whole and not just declarative pointing) predicted later receptive vocabulary on the PPVT, whereas Reach-request gestures were negatively related to MLU and Protest gestures to communicative competence. While the correlations are moderate, the long-term relationships (over more than 20 months) are impressive. They also show the importance of studying the whole repertoire to determine not only positive relationships but also negative ones. Not only do object-sharing and information-sharing gestures apparently promote verbal communication, but also primitive demanding and rejecting gestures appear to interfere with it. In the Italian-Canadian sample, female children were higher than males in vocabulary size at 15 months and MLU at 3 years, but not in PPVT or communicative competence (Blake et al., 1997). There were no gender differences at 15 months in Object exchange or Reach-request gestures.

Our results support those studies previously cited that found an increase in pointing in the second year. They do not, however, support those that found a specific relationship between declarative pointing and language measures. Butterworth and Morissette (1996) also did not find a relationship between declarative pointing (specifically, age of onset) and language (vocabulary production). Rather, we found a relationship between Comment gestures as a whole and receptive vocabulary. The fact that comprehension rather than production is predicted by deictic gestures agrees with the findings of Bates et al. (1989), but their deictic gestures included also giving and reaching. That Agency gestures were also related to vocabulary supports the findings of Camaioni et al. (1991); but their results relate such gestures to productive vocabulary, whereas ours relate them to receptive vocabulary.

In conclusion, the studies reviewed reveal great similarity across the samples compared in the types of gestures that increase or decrease over the transition period to language. They also partially supported the prediction that certain types of gestures would be positively related to language measures, while others would be negatively related. The advantages of examining the entire repertoire of bodily gestures in infants are apparent in that we can determine those aspects of nonverbal communication that are beneficial for language and those that are not. This information should be useful for intervention studies with preverbal handicapped infants.

Acknowledgements

The research was supported by grants to the first author from the Natural Sciences and Engineering Council of Canada, from the Social Sciences and Humanities Research Council Grant to York University, from the Faculty of Arts, York University, from the Elia Research fund, and by a fellowship from the Fondation Fyssen to the first author. We are grateful to the participating families in all countries for allowing our many visits. We would also like to thank Sheila McConnell for her help in devising the coding system, Mrs. Naoko Ito for conducting the home visits in Japan, the late Nancy Benson for providing the videotapes of three of the English-Canadian infants, Silvana Macdonald for analyzing the measure of communicative competence, and Mythili Viswanathan for the Figures.

References

Bates, Elizabeth, Laura Benigni, Inge Bretherton, Luiga Camaioni, & Virginia Volterra (1979). *The emergence of symbols: Cognition and communication in infancy.* New York: Academic Press.

Bates, Elizabeth, Donna Thal, Kimberly Whitesell, Larry Fenson, & Lisa Oakes (1989). Integrating language and gesture in infancy. *Developmental Psychology, 25,* 1004–1019.

Blake, Joanna (2000). *Routes to child language: Evolutionary and developmental precursors.* NY: Cambridge University Press.

Blake, Joanna & Susan Dolgoy (1993). Gestural development and its relation to cognition during the transition to language. *Journal of Nonverbal Behavior, 17,* 87–102.

Blake, Joanna, Esther Olshansky, Grace Vitale, & Silvana Macdonald (1997). Are communicative gestures the substrate of language? *Evolution of Communication, 1,* 261–282.

Blake, Joanna, Patricia Osborne, Marlene Cabral, & Pamela Gluck (2003). The development of communicative gestures in Japanese infants. *First Language, 23,* 3–20.

Blake, Joanna, Georgia Quartaro, & Susan Onorati (1993). Evaluating quantitative measures of grammatical complexity in spontaneous speech samples. *Journal of Child Language, 20,* 139–152.

Butterworth, George & Paul Morissette (1996). Onset of pointing and the acquisition of language in infancy. *Journal of Reproductive and Infant Psychology, 14,* 219–231.

Camaioni, Luiga, M. Cristina Caselli, Emiddia Longobardi, & Virginia Volterra (1991). A parent report instrument for early language assessment. *First Language, 11,* 345–359.

Camaioni, Luiga, Paola Perucchini, Francesca Bellagamba, & Cristina Colonnesi (2004). The role of declarative pointing in developing a theory of mind. *Infancy, 5,* 291–308.

Capirci, Olga, Jana M. Iverson, Sandro Montanari, & Virginia Volterra (2002). Gestural, signed and spoken modalities in early language development: The role of linguistic input. *Bilingualism: Language and Cognition, 5,* 25–37.

Carpenter, Malinda, Katherine Nagell, & Michael Tomasello (1998). Social cognition, joint attention and communicative competence from 9 to 15 months of age. *Monographs of the Society for Research in Child Development, 63* (Serial No. 255).

Cheek, Adrianne, Kearsy Cormier, Ann Repp, & Richard P. Meier (2001). Prelinguistic gesture predicts mastery and error in the production of early signs. *Language, 77,* 292–323.

Desrochers, Stéphan, Paul Morisette, & Marcelle Ricard (1995). Two perspectives on pointing in infancy. In Chris Moore & Philip J. Dunham (Eds.), *Joint attention: Its origins and role in development* (pp. 85–101). Hillsdale, NJ: Lawrence Erlbaum.

Dunn, Lloyd M. & Leota M. Dunn (1981). *The Peabody Picture Vocabulary Test.* Circle Pines, MN: American Guidance Service.

Ekman, Paul & Wallace Friesen (1969). The repertoire of nonverbal behavior: Categories, origins, usage, and coding. *Semiotica, 1,* 49–97.

Guidetti, Michèle (2002). The emergence of pragmatics: Forms and functions of conventional gestures in young French children. *First Language, 22,* 265–285.

Iverson, Jana M., Olga Capirci, & M. Cristina Caselli (1994). From communication to language in two modalities. *Cognitive Development, 9,* 23–43.

Iverson, Jana M. & Donna J. Thal (1998). Communicative transitions: There's more to the hand than meets the eye. In Amy M. Wetherby, Steven F. Warren, & Joe Reichle (Eds.), *Transitions in prelinguistic communication,* Vol. 7. (pp. 59–86). Baltimore: Brookes.

Laakso, M.-L., A.-M. Poikkeus, J. Katajamaki, & P. Lyytinen (1999). Early intentional communication as a predictor of language development in young toddlers. *First Language, 19,* 207–231.

Legerstee, Maria & Yarixa Barillas (2003). Sharing attention and pointing to objects at 12 months: Is the intentional stance implied? *Cognitive Development, 18,* 91–110.

Lempers, Jacques D. (1979). Young children's production and comprehension of nonverbal deictic behaviors. *The Journal of Genetic Psychology, 135,* 93–102.

Leung, Eleanor H.L. & Harriet L. Rheingold (1981). Development of pointing as a social gesture. *Developmental Psychology, 17,* 215–220.

Liszkowski, Ulf, Malinda Carpenter, Anne Henning, Tricia Striano, & Michael Tomasello (2004). Twelve-month-olds point to share attention and interest. *Developmental Science, 7,* 297–307.

Meltzoff, Andrew N. (1995). Understanding the intentions of others: Re-enactment of intended acts by 18-month-old children. *Developmental Psychology, 31,* 838–850.

Murphy, Catherine M. & David J. Messer (1977). Mothers, infants and pointing: A study of a gesture. In H. Rudolph Schaffer (Ed.), *Studies in mother-infant interaction* (pp. 325–354). London: Academic Press.

Ogura, Tamiko & Toru Watamaki (1998). Research and development of early language development inventory. Research Project, Grant-in-aid for Scientific Research. Kobe University.

Ogura, Tamiko & Toru Watamaki (2001). The relationship between gesture and language development through the Japanese version of the CDI. Paper presented at the Biennial Meeting of the Society for Research in Child Development, April.

Rodrigo, Maria José, Angela González, Manuel de Vega, Mercedes Muñetón-Ayala, & Guacimara Rodriguez (2004). From gestural to verbal deixis: A longitudinal study with Spanish infants and toddlers. *First Language, 24,* 71–90.

Rowe, Meredith L. (2000). Pointing and talk by low-income mothers and their 14-month-old children. *First Language, 20,* 305–330.

Wilkins, David (2003). Why pointing with the index finger is not a universal (in sociocultural and semiotic terms). In Sotaro Kita (Ed.), *Pointing: Where language, culture and cognition meet* (pp. 171–215). Mahwah, N.J.: Lawrence Erlbaum.

Zinober, Brenda & Margaret Martlew (1985). Developmental change in four types of gesture in relation to acts and vocalizations from 10–21 months. *British Journal of* Developmental Psychology, 3, 293–306.

How does linguistic framing of events influence co-speech gestures?

Insights from crosslinguistic variations and similarities

Asli Özyürek[1,2], Sotaro Kita[3], Shanley Allen[4], Reyhan Furman[5], and Amanda Brown[2,4]
[1]F.C. Donders Centre for Cognitive Neuroimaging, Radboud University / [2]Max Planck Institute for Psycholinguistics, Nijmegen / [3]University of Bristol, UK / [4]Boston University/ [5]Bogaziçi University, Istanbul

What are the relations between linguistic encoding and gestural representations of events during online speaking? The few studies that have been conducted on this topic have yielded somewhat incompatible results with regard to whether and how gestural representations of events change with differences in the preferred semantic and syntactic encoding possibilities of languages. Here we provide large scale semantic, syntactic and temporal analyses of speech- gesture pairs that depict 10 different motion events from 20 Turkish and 20 English speakers. We find that the gestural representations of the same events differ across languages when they are encoded by different syntactic frames (i.e., verb-framed or satellite-framed). However, where there are similarities across languages, such as omission of a certain element of the event in the linguistic encoding, gestural representations also look similar and omit the same content. The results are discussed in terms of what gestures reveal about the influence of language specific encoding on on-line thinking patterns and the underlying interactions between speech and gesture during the speaking process.

Introduction

When we talk about events that we have observed, we segment our continuous perception into units that are verbalizable. All languages of the world (spoken or signed) have such properties, that is, lexical elements that segment aspects of an event and sequence them in a hierarchical fashion following certain linguistic constraints. For example, to talk about someone running across the street, one has to use separate lexical items to express the figure, manner and path of motion as well as the ground on which the motion takes place, as in the sentence "the man ran across the street".

However, languages differ with regard to which semantic elements comprising the events are readily encoded and lexicalized. For example, while German and Dutch verbs of placing readily encode the position of the object placed in the semantics of the verb (e.g., Dutch: *leggen* for horizontal placement versus *zetten* for vertical placement) (Lemmens, 2002), English does not have separate verbs that encode such distinctions, even though they could be paraphrased in several ways when necessary (i.e., *put the book on the shelf in a vertical position*) (also see Levinson & Meira, 2003, for an overview of such differences across many languages).

Another variation in event descriptions across languages concerns the way semantic elements of an event are mapped onto syntactic structures (e.g., Slobin, 1987; Talmy, 1985). For example, different languages can be classified as either "satellite-framed" or "verb-framed" depending on how path and manner components of a motion event are typically lexicalized (Talmy, 1985). Speakers of satellite-framed languages such as English express the path in a so-called satellite, like *up*, and manner in the main verb, as shown in (1). However, in verb-framed languages such as Turkish, Spanish etc., path information is expressed in the main verb and manner information outside of it. In Romance languages like Spanish, manner is frequently expressed in the form of an adverbial gerund (2) (Slobin, 1996). In other verb-framed languages such as Turkish or Japanese, manner is expressed typically in the verb of the subordinate clause rather than as an adverbial gerund, as in the Turkish sentence in (3) (Kita & Özyürek, 2003; Özyürek & Kita, 1999).

(1) *The ball rolled down the hill*

(2) *Sale volando*
Exit fly-Gerund
He/she/it exits flying

(3) *Top zıplay-arak tepe-den aşağı in-di.*
ball jump-Connective hill-Ablative downness descend-Past
'The ball descended the hill while jumping.'

What about co-speech gestures that we use as we talk about such elements of events? Co-speech gestures are spontaneous and frequent accompaniments to speech and the expressions in the two modalities have been found to be tightly integrated pragmatically, semantically, and temporally (Clark, 1996; Goldin-Meadow, 2004; Kendon, 2004; McNeill, 1985, 1992). A subset of co-speech gestures that are frequently used in event descriptions are called *iconic* gestures (McNeill, 1992) which convey meaning by their "iconic" resemblance to the different aspects of the events they depict (e.g., wiggling fingers crossing space to represent someone walking).

The question we address in the present study is whether iconic gestures are influenced by the semantic and syntactic encoding of aspects of events during online speaking and how such influence is realized. Iconic gestures are very appropriate to investigate such a question since in their representations they involve semantic elements of event components such as figure, ground, manner and path that are also encoded by languages in different ways. Thus they allow to investigate the question of whether co-speech gestures represent elements of the events as imagined (i.e., with no influence from linguistic representations of events) or are shaped by the way event components are represented by the syntactic and semantic structures of different languages. Recent research has provided evidence for the latter claim (Kita, 2000; Kita & Özyürek, 2003; McNeill, 2000; McNeill & Duncan, 2000; Müller, 1998) showing that gestures of the same events differ across speakers of typologically different languages. These findings are compatible with the view that in utterance generation there is interaction between linguistic and gestural/imagistic representations of events (e.g., Interface Hypothesis: Kita & Özyürek, 2003; Growth Point theory: McNeill & Duncan, 2000)

However, in the literature, the type of semantic coordination between gestural and linguistic representations have been depicted in different ways. For example, McNeill and Duncan compared English and Spanish speakers' depictions of motion events in narrations of a Sylvester and Tweety cartoon, "Canary Row" (McNeill, 2000; McNeill & Duncan, 2000). They found that Spanish speakers are more likely to omit manner in their spoken utterances about a motion event that includes manner — possibly due to the fact that manner is not encoded in the main verb and that manner verbs are not very rich in Spanish (Slobin, 1996). In such cases, Spanish speakers have been found to express manner in a *compensatory* way in their gesture and distribute the gesture over many path phrases in speech (i.e., "manner fog"), as illustrated by individual examples. In contrast, English speakers are more likely to express manner together with path in their speech. Their gestures have been found to either express both manner and path or downplay the manner and show only path in certain discourse contexts- i.e., when manner is not the new or focus of information. Thus there is evidence that gestural representations of events compensate for those expressed in speech and provide extra information in languages where speakers are more likely to omit certain elements than other languages (i.e., Spanish versus English).

However, Kita and Özyürek (Kita & Özyürek, 2003; Özyürek & Kita, 1999) have shown other types of semantic coordination between gestural and linguistic representations of events. They compared how Japanese, Turkish, and English speakers verbally and gesturally express motion events with data taken from the narrations of the above-mentioned animated cartoon. In their studies two analyses were carried out. The first analysis concerned an event in which a protagonist swung on a rope like Tarzan from one building to another. In Turkish and Japanese, there is no readily accessible expression that semantically encodes agentive change of location of the protagonist with an arc trajectory. Whereas in English, this aspect of the event can be easily encoded with the verb *to swing*. Thus it was found that all English speakers used the verb *swing*, which encodes the arc shape of the trajectory. On the other hand, Japanese and Turkish speakers used verbs which do not encode the arc trajectory (e.g., Turkish: *gidiyor* (goes)). Paralleling these distinctions in semantic encoding, it was found that Japanese and Turkish speakers were more likely to produce a straight gesture, which does not encode the arc trajectory, than English speakers and English speakers produced only gestures with an arc trajectory.

Their second analysis concerned how speech and gesture express the manner and path of an event in which the protagonist rolled down a hill. It was found that verbal descriptions differed cross-linguistically in terms of how manner and path information is lexicalized along the lines discussed by Talmy (1985). English speakers typically used a manner verb and a path particle or preposition (e.g., *he rolled down the hill*) to express the two pieces information within one clause. In contrast, Japanese and Turkish speakers separated manner and path expressions over two clauses with path in the main clause and manner in the subordinated clause (e.g., *he descended as he rolled*). Given the assumption that a clause approximates a unit of processing in speech production (e.g., Garrett, 1982; Levelt, 1989), presumably English speakers were likely to process both manner and path within a single processing unit, whereas Japanese and Turkish speakers were likely to need two processing units. Consequently, Japanese and Turkish speakers should be more likely than English speakers to separate the components of manner and path in preparation for speaking so that the two pieces of information could be dealt with in turn, as compared to English speakers. The gesture data confirmed this prediction. In depicting how an animated figure rolled down a hill after having swallowed a bowling ball, Japanese and Turkish speakers were more likely to use separate gestures, one for manner and one for path, and English speakers were indeed more likely to use just one gesture to express both. It was concluded from these two sets of findings that gestures encode those aspects that fit conceptualization of the event for the purposes of speaking in a particular language, rather than compensate for aspects of representations that are hard to encode linguistically or omitted for discourse purposes.[1]

However, the few studies conducted on this topic have certain limitations. Some were based on individual examples of speech and gesture pairs without providing quantitative distribution from a variety of speakers of each language (McNeill & Duncan, 2000). Others were conducted on only a few motion event types and without taking the tight temporal synchrony between speech and gestures into account (Kita & Özyürek, 2003; Özyürek & Kita, 1999). In addition, the possibility could not be ruled out that the gestural differences across languages could be due to other factors (e.g., cultural differences in patterns of movement, preference for certain perspectives in depicting events, etc.) than differences in the syntactic or lexical semantic encoding of the events per se.

In the present study, we try to overcome the limitations of previous studies by investigating speech and gesture relations in 10 different motion event descriptions from 20 English and 20 Turkish speakers. In our analyses we compare different types of linguistic framing of events and the co-occurring gestures. First, we investigate the nature of gestural representations when speakers encode one element of the event and omit the other in their speech to see whether gestures co-occurring with these utterances compensate for the omitted information or not. Secondly, we compare gestural representations when speakers express both manner and path in their speech but differ in the way this information is encoded syntactically, that is, in one verbal clause (satellite-frame, in English) or two verbal clauses (verb-frame, in Turkish) across languages. Finally, we also analyze to what extent the expressions of manner and/or path information in gestures synchronize in time with expressions of manner and path information in speech and whether languages differ in this respect or not.

According to Interface Hypothesis of speech and gesture production (Kita & Özyürek, 2003), there is interaction between linguistic and gestural/imagistic representations of events during speaking. That is, gestures do not merely encode imagistic representations of the event but those aspects that fit conceptualization of the event for the purposes of speaking. Gestures are shaped by the linguistic encoding choices and representations during speaking. Thus according to this hypothesis, when speakers omit one of the event components in their speech, then we expect them to omit that component in their gestures as well, regardless of the language spoken. However, in cases where speakers of languages express both event components but use different syntactic frames to do so, then we expect their gestures to also differ in order to fit the conceptualization of the event for speaking. If these predictions hold, then we argue gestures to be revealing *thinking-for-speaking* (Slobin, 1987, 1996) patterns, that is, language specific conceptualization and thinking patterns tuned and adopted for linguistic encoding choices rather than merely the spatial imagery of the event.

Method

Participants

Participants in the study were 20 monolingual Turkish and 20 monolingual English speakers. All were adults ranging in age from about 18 to 40 and were university students in either Istanbul (Turkish) or Boston (English).

Materials

Data were collected by elicitation, using a set of ten video clips depicting motion events involving simultaneous manner and path (Özyürek, Kita, & Allen, 2001). Five manners and three paths were depicted, yielding the following combinations: JUMP+ASCEND, JUMP+DESCEND, JUMP+GO.AROUND, ROLL+ASCEND, ROLL+DESCEND, ROTATE+ASCEND, ROTATE+DESCEND, SPIN+ASCEND, SPIN+DESCEND, and TUMBLE+DESCEND. The manner JUMP involves an object moving vertically up and down (moving along a flat or inclined surface), ROLL involves an object turning on its horizontal axis (moving along an inclined surface), ROTATE and TUMBLE both involve an object turning on its horizontal axis (moving vertically through the air), and SPIN involves an object turning on its vertical axis (moving along an inclined surface).

Each video clip was between 6 and 15 seconds in duration, and had three salient components: an entry event, a target motion event, and a closing event. All clips involved a round red smiling character and a triangular-shaped green frowning character, moving in a simple landscape. We refer to them here as Tomato Man and Green Man. As an example, the ROLL+DESCEND clip goes as follows. The initial landscape on the screen is a large hill ending in a tree; Tomato Man is located at the top of the hill. Green Man enters the scene from the right and bumps into Tomato Man [entry event], then Tomato Man rolls down the hill [target motion event], and finally hits the tree [closing event], as illustrated in Figure 1.

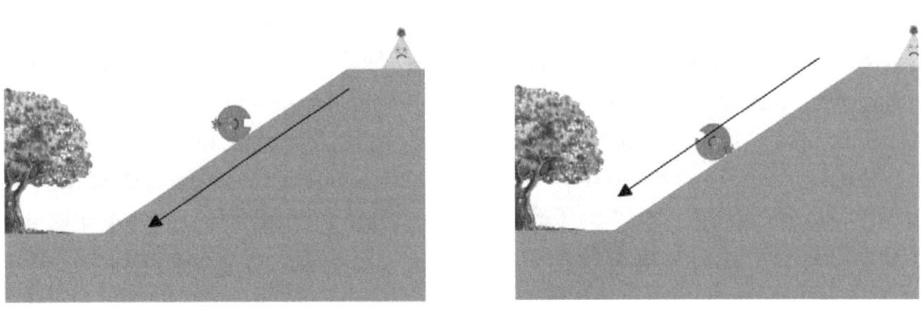

Figure 1. Selected stills of target event ROLL+DESCEND

Procedure

Participants were tested individually in a quiet space at their university. All interactions were videotaped for later coding and analysis. During the warm-up phase, the experimenter showed participants a typical scene from a clip and introduced them to the characters and the landscape. She then gave them two practice rounds with clips involving motion events like those in the test clips and asked them to tell what happened in the clip to a listener who purportedly had not seen it.

In the testing phase, the experimenter presented the ten test clips for the participant to narrate, following the same format as in the warm-up phase. If participants did not mention the target event in their narration, either the experimenter or the listener encouraged them to do so with a question that did not focus explicitly on either manner or path.

Transcription and Coding

Speech Transcription

All speech relevant to the target motion events was transcribed by native speakers of the relevant language into MediaTagger, a video-based computer program (Brugman & Kita, 1995). Note that we did not transcribe any speech that described exclusively the entry event, the closing event, or the setting of the scene. The relevant speech for each participant was segmented into "sentences," which we define here as a matrix clause plus its subordinates, if any. Examples are shown in (4) and (5), with sentence segmentations indicated by square brackets.

(4) English:
  a. [*He rolled up the hill.*]
  b. [*And he is spinning as he goes down the hill.*]
  Turkish:
  c. [*Domates adam yuvarlan-arak tepe-yi çıkı-yo*]
  tomato man roll-Connective hill-Accusative ascend-Present
  'tomato man ascends the hill while rolling'

Two matrix clauses separated by a coordinating conjunction (i.e., *and, but,* and *or* for English, and *ve, sonra* for Turkish) were considered as two sentences. Many participants used more than one sentence to describe a given target motion event. We refer to the full set of sentences that describe a particular target motion event as a "target-event description."

The description for the ROTATE+DESCEND clip given in (5) illustrates this process. The speaker uttered everything in (5) as his description of the clip. Only the portion describing the target motion event of the Tomato Man rotating down into the water (i.e., the target-event description) was transcribed into MediaTagger, as indicated in (5) by curly brackets. The target-event description was then divided into three sentences, as indicated by square brackets.

(5) *There's a ledge on the right and Triangle Man is floating in the water on the left. Tomato Man slides off sort of Wile E. Coyote style, where he doesn't just fall straight off, but goes about halfway in the air* {[*and then falls down*]. [*So he spins down,*] [*spins down*]} *and lands next to Triangle Man.*

In order to establish reliability of the identification and segmentation of sentences, twenty percent of the data were independently processed by a second coder who was either a fluent or native speaker of the relevant language. For each clip, the second coder identified the stretch of discourse describing the target event and segmented it into sentences. The percentages of the original coder's sentences with which the second coder agreed in terms of identification and segmentation of target event sentences were as follows: for English, 92% and for Turkish, 88%.

Speech Coding

Each sentence was coded by native speakers of the relevant language as one of four categories according to the structural patterns of information packaging in speech relating to manner and path. In this coding, manner refers to the secondary movement (rotation along different axes, or jumping) of the figure that co-occurs with the translational movement in the target events. Path refers to the directionality or trajectory specifications for the translational movement.

Some sentences included only one of the motion elements such as manner or path. The first category, "Manner-only," denotes use of only a manner element in the sentence (i.e., no path). Sentences coded as Manner-only in English include simple manner verbs (6a), manner verbs with or without some further description of the manner, and phrases which describe the manner without a manner verb (6b).

(6) a. *And then tumbles head over heels.*
b. *And does a little couple of rounds.*

Turkish Manner-only sentences include constructions similar to the English ones shown in (6a), but nothing like that in (6b).

The next category, "Path-only," indicates use of only a path element (i.e. no manner) in the sentence. In English, sentences coded as Path-only include the light path verb *go* followed by directional path particles or adpositional phrases (7a), or other path verbs optionally followed by directional path particles or adpositional phrases (7b).

(7) a. *He goes up a hill.*
b. *It fell.*

In Turkish, sentences coded as Path-only include light path verbs (*come* and *go*) as in (8a) and other path verbs as in (8b), both with optional postpositional phrases that include spatial nouns specifying the source or the goal of the path.

(8) a. *Aşağı-ya gel-iyor.*
downness-Dative come-Present
'(He/she/it) comes down.'
b. *Sonra yukarı çık-tı.*
then upness ascend-Past
'Then (he/she/it) ascended (to) the upness.'

For sentences in which both manner and path were mentioned, two coding categories were distinguished. The category "Tight" denotes a tight packaging of both manner and path in one sentence, that is, a unit involving one verb and one closely associated non-verbal phrase. Sentences coded as Tight differ somewhat across languages. English Tight sentences include manner verbs followed by directional path particles or prepositional phrases (9a), and phrases describing manner followed by a directional path particle or prepositional phrase (9b).

(9) a. *He rolled up the hill.*
b. *And he did his little two-step down the hill.*

Tight sentences also occur in Turkish, although they were rarely used. A typical example of this includes a manner verb with a postpositional directional path phrase, but crucially no path verb (10).

(10) *Domates adam aşağı yuvarlan-ıyor tepe-den.*
tomato man downness roll-Present hill-Ablative
'Tomato Man rolls down the hill.'

The second category of sentences in which both manner and path were mentioned is labeled "Semi-tight." This code denotes a semi-tight packaging of manner and path in one sentence, with each of these expressed by a separate verbal element, one subordinated to the other. In English, the subordinated form can be either a fully tensed verb (11a) or a progressive participle functioning as an adverbial (11b).

(11) a. *He spins in circles while he's going down.*
b. *Triangle Man ascends the hill twirling.*

In Semi-tight constructions in Turkish, the manner verb is subordinated to the main path verb with the use of a connective — mostly *-arak* as in (12a), and very rarely *–ip*. Another possibility is to use a reduplicated manner verb functioning as an adverbial and subordinated to the main path verb, as in (12b).

(12) a. *Domates adam yuvarlan-arak yokuş-u in-di.*
tomato man roll-Connective hill-Accusative descend-Past
'Tomato man descended the hill while rolling.'
b. *Űçgen dőne-dőne çık-tı.*
triangle turning-turning ascend-Past
'Triangle ascended turning turning.'

Finally, sentences which included more than one type of packaging of manner and path were coded for each relevant type. For example, the sentence in (13) was coded as both Path-only and Tight.

(13) *When he went down, he was spinning down.*

In order to establish reliability of the coding, twenty percent of the data were independently processed by a second coder who was either a fluent or native speaker of the relevant language. The second coder judged the category type (i.e., Manner-only, Path-only, Tight, Semi-tight) for each sentence that had been segmented and transcribed by the original coder. The agreement between coders for this judgment was as follows: for English: 93%, and for Turkish: 91%.

Gesture transcription
We transcribed gestures that occurred concurrent with sentences in the target event descriptions that contained manner and/or path. The stroke phase of gestures (Kendon, 1980; McNeill, 1992) was isolated by frame by frame video analysis, according to the procedure detailed in Kita, van Gijn, and van der Hulst (1998). We excluded gestures that overlapped with more than one utterance type (e.g., Tight and Path only) or non target-event utterances (see speech transcription above).

Gesture coding
Gestures that encoded the manner and/or path of the target event were called *target-event gestures*. They were classified into five types: *Manner*, *Path*, *Conflated*, *Combined or Unclear*. Manner gestures encoded manner of motion (e.g., a repetitive up and down movement of the hand to represent jumping) without encoding path. Path gestures expressed change of location without encoding manner. Conflated gestures expressed both manner and path at the same time all throughout the stroke (e.g., repetitive up and down movements superimposed on diagonal downward change of location of the hand, representing jumping down the slope).

Combined gestures were two-handed gestures, in which each hand was a different type (e.g., one hand was Manner, the other hand was Conflated). Finally some gestures were coded as Unclear either because they were hard to segment, hard to categorize for any of the two categories above or were unclear with regard to whether they were representational gestures or just self-adaptors.

For purposes of clarity we excluded gestures that were Combined or Unclear from the analysis. We also excluded the few gestures that included the use of body to represent change of location or manner (e.g., trying to show rotation using mainly head, shoulders and arms) since these representations bias towards Manner only and thus do not allow comparison for the frequency of the use of Conflated gestures.

In order to establish reliability of the identification and segmentation of target-event gestures, twenty percent of the data was independently processed by a second coder. For each clip, the second coder identified target-event gestures in the discourse, and segmented the stroke phase of the gestures. 81% of the original coder's gesture strokes (N = 108) had an overlap with a gesture stroke identified and segmented by the second coder. Among these gestures, the discrepancy between the two coders was on average 1.72 frames ($SD = 2.02$) (1 frame = 33.3ms) at the beginning of the stroke and 2.54 frames ($SD = 4.74$) at the end. Among the gesture strokes that were identified by both coders, 90% of the original coder's strokes overlapped with a stroke coded by the second coder coded with the discrepancy of 5 video frames (167ms) or less at the beginning and the end of the stroke.

Furthermore, in order to establish reliability of the gesture type classification, the second coder judged the gesture type (i.e., Manner, Path, Conflated, etc.) for each target-event gesture stroke that had been identified and segmented by the original coder. The agreement between coders was 89% for the gesture type.

Results

Speech

Since our main investigation is to see whether gestural representations of the same events differ with different syntactic constructions across languages and how, first we analyzed the differences in speech between the languages. First, we investigated to what extent speakers of different languages were more likely to mention only one of the elements (i.e., either path or manner) in their sentences about the target events. Secondly, we focused on only those sentences in which both manner and path were expressed and compared the type of syntactic packaging (i.e., Tight or Semi-tight) that each group of participants used in their descriptions.

Inclusion of manner and/or path across languages: The first analysis showed that the proportion of sentences that included both types of information (i.e., path and manner) was similar across languages; English, $M = 0.62$, $SD = 0.16$, and Turkish, $M = 0.65$, SD = 0.19. In the rest of the sentences either Path or Manner was omitted. To compare whether languages differed in terms of using Path-only or Manner-only sentences, a 2*2 repeated measures ANOVA with language type (Turkish and English) as the between subjects variable and information type (Manner-only or Path-only) as the within subjects variable was conducted. There was a main effect of information type (F (1,38) = 50.991; $p < 0.001$), but no interaction with language. This showed that both Turkish and English speakers used a similar amount of Path-only or Manner-only sentences (see Table 1). However, all speakers preferred Path-only sentences to Manner-only sentences, regardless of the language they were speaking, as evidenced by the main effect.

Differences in syntactic packaging between the languages: In the next analysis the proportion of sentences in which both manner and path are expressed were compared in terms of the type of syntactic packaging preferred in each of the two languages. A 2*2 repeated measures ANOVA with language (Turkish versus English) as the between subjects variable and syntactic packaging type (Tight versus Semi-tight) as the within subjects variable revealed an interaction between language and syntactic type (F (1,38) = 269.69, $p < 0.001$). As expected from the typological differences, further Bonferroni adjusted t tests revealed that English speakers used more Tight syntactic packaging than Turkish speakers (t (38) = −13.4, $p < 0.001$) and Turkish speakers used more Semi-tight packaging than English speakers (t (38) = 11.76, $p < 0.001$) (see Table 1).

To summarize, the speech analysis showed that the speakers' choices of focusing on one element of the event was similar across the languages. However, when they expressed both components, English and Turkish speakers used different syntactic framing for expressing event components. Next we investigated how gestural representations co-occurred with these different linguistic choices.

Table 1. Average proportion of sentences that express different types of manner and path information across languages

	Both Manner and Path	**Manner Only**	**Path Only**	**Total N of sentences**
English	0.62 (0.16)	0.12 (0.09)	0.26 (0.12)	315
Tight	0.52 (0.16)			
Semi-tight	0.07 (0.07)			
Turkish	0.65 (0.18)	0.07 (0.08)	0.28 (0.14)	294
Tight	0.02 (0.01)			
Semi–tight	0.63 (0.19)			

Note: Parentheses indicate standard deviations

Gesture

In the gesture analysis, we first looked at how gestures represented information when either the path or the manner element was omitted from the sentence. Secondly, we analyzed whether the differences in syntactic packaging of information influenced the information packaging of information in gesture when both manner and path were expressed in speech.

All the analyses were conducted on the proportions of gesture types out of all the gestures that co-occurred temporally with different sentence types (e.g., Path-only, Manner-only, Tight, Semi-tight etc.) for each subject.

Relations between the informational content of gestures and speech: One crucial question is when speakers express only one piece of information in their sentences (i.e., in Manner-only and Path-only sentences), what does the content of their gestures look like? Do they compensate for the omission of information in speech or do they also omit that same information in gestures? To investigate this question, we analyzed the type of gestures that accompanied Manner-only and Path-only sentences.

Gestures in the context of Path-only sentences: In this analysis, only gestures that accompanied Path-only sentences were included. Each subject had to contribute with at least 5 gestures for his/her data to be included in this analysis to have enough variance for the statistical comparisons. A total of 11 English and 7 Turkish speakers were included in the analysis. The proportions of each type of gesture out of all the gestures that occurred during Path-only sentences were calculated (see Figure 2). A 2*3 repeated measures ANOVA with language as the between subjects variable and gesture type (Manner, Path or Conflated) as the within subjects variable revealed only an effect of gesture type (F (2,32) = 217.07; $p < 0.001$). That is, speakers of English and Turkish did not differ in terms of the type of gesture they preferred when they expressed only path information in speech. Further Bonferroni adjusted t tests among the gesture types revealed that speakers of both languages preferred Path over Manner (t (17) = −18.7, $p < 0.001$) or Conflated gestures (t (17) = −14.07, $p < 0.001$) when they expressed only path information in their speech.

Gestures in the context of Manner-only sentences: In this analysis, only gestures that accompanied Manner-only sentences were included. Due to the fact that there were many fewer Manner-only sentences, each subject had to contribute at least 3 gestures to be included in this analysis. A total of 8 English speakers and 4 Turkish speakers were included in the analysis. A 2*3 repeated measures ANOVA with language as the between subjects variable and gesture type (Manner, Path or Conflated) as the within subject variable revealed only an effect of gesture type (F (2,20) = 4.58; $p < 0.05$). That is, speakers of English and Turkish did not differ in

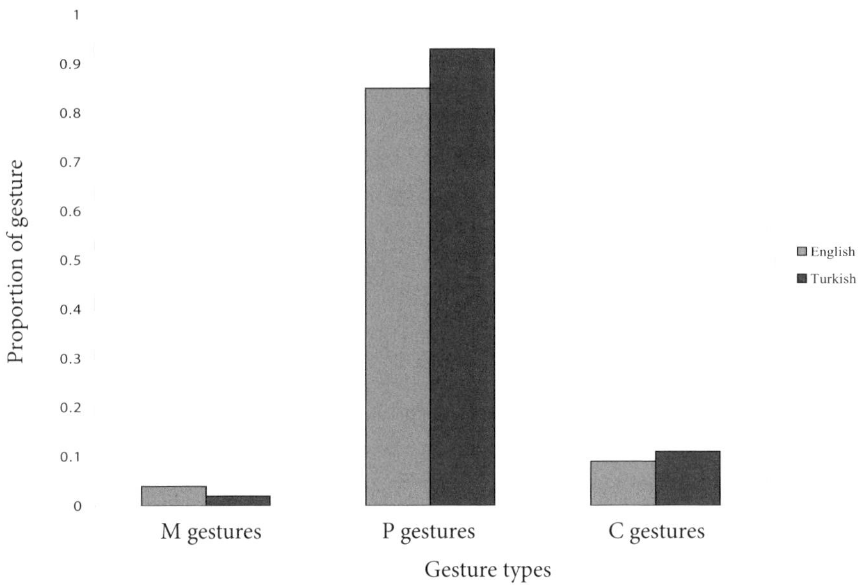

Figure 2. Proportions of gesture types (Manner, Path, Conflated) accompanying Path-only sentences across languages

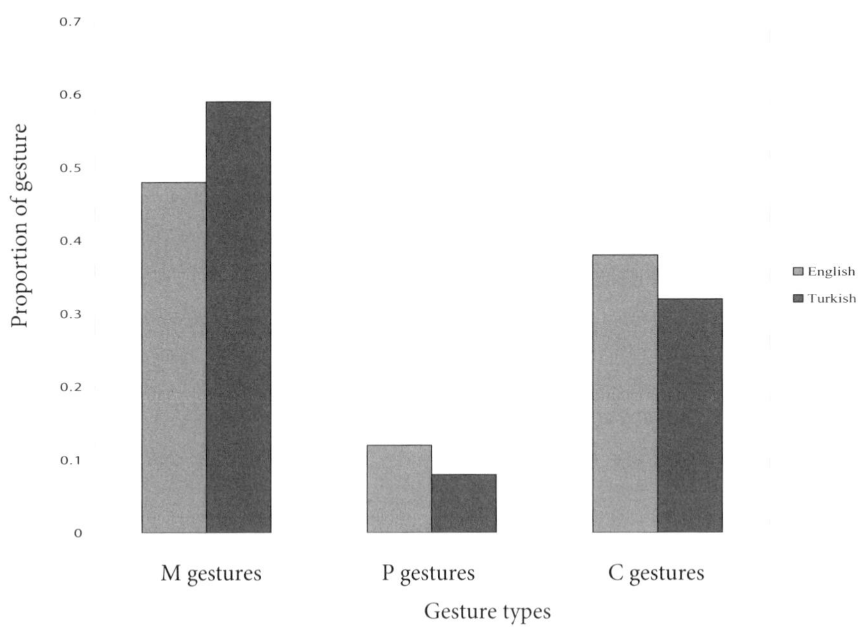

Figure 3. Proportions of gesture types (Manner, Path, Conflated) accompanying Manner-only sentences across languages

terms of the type of gesture they preferred when they expressed only manner information in speech. Further Bonferroni adjusted *t* tests among the gestures types revealed that speakers of both languages used more Manner (t (11) = 3.40; $p < 0.01$) and Conflated (t (11) = 2.79; $p < 0.05$) gestures than Path gestures when they expressed only manner information in their sentences (see Figure 3).

In sum, the analysis showed that the information expressed both in gesture and speech showed strong parallels regardless of the language spoken. That is, when speakers of both languages expressed only path in their speech they were more likely to use Path gestures. Likewise when they expressed only manner in their speech, they included gestures that contained manner (both Manner and Conflated gestures), but crucially not Path gestures that would mismatch, or compensate the informational content of the utterance.

Gestures accompanying different syntactic packaging of manner and path information: In this analysis, only gestures that accompanied Tight sentences in the English sample and Semi-tight sentences in the Turkish sample were included. Each subject had to contribute at least 5 gestures to be included in this analysis. A total of 18 English speakers and 20 Turkish speakers were included in the analysis. A 2*3 repeated measures ANOVA with language (Turkish and English) as the between subjects variable and the gesture type (Manner, Path or Conflated) as the within subject variable revealed interaction between the two factors (F (2,72) = 19.33; $p < 0.001$). Bonferroni adjusted t tests showed that English speakers used more Conflated gestures than Turkish speakers (t (36) = 5.55; $p < 0.001$). On the other hand Turkish speakers used more Manner (t (36) = –3.14; $p < 0.05$) and Path (t (36) = –3.4, $p < 0.05$) gestures than English speakers (see Figure 4).

This analysis shows that in addition to the informational content, syntactic packaging of information also influences the type of gestural representations. Namely, gestures reveal similar representations to the linguistic encoding of events.

Temporal synchrony of information between gestures and speech: In the final analysis we investigated the tight temporal synchrony between the types of gestures and element of speech they co-occurred with in the sentence. If gestures are influenced by the online conceptualization of events during online syntactic and semantic encoding, we expect them also to be tightly synchronized with the speech segment they co-occur with. We tested this hypothesis with Manner and Path gestures since Conflated gestures could overlap with speech that could express both manner and path. We calculated the proportion of Path gestures that co-occurred with path speech (i.e., with path verbs, path particles, and prepositional phrases). Note that path gestures were not included in the "path matching" category if they co-occurred exclusively with manner speech, expression of figure (e.g., tomato man) or discourse markers, thus without overlapping with any path

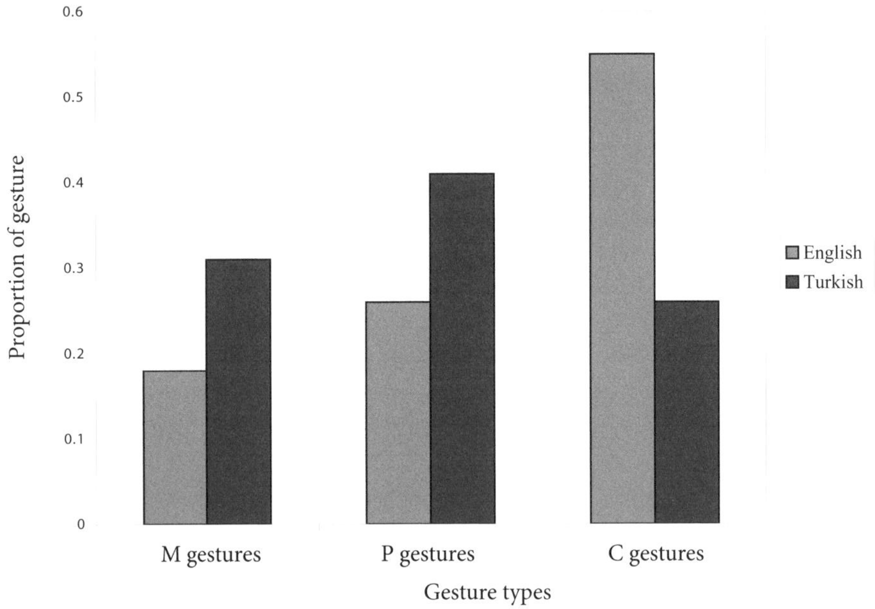

Figure 4. Proportions of gesture types (Manner, Path, Conflated) accompanying Tight and Semi-tight sentences across languages

speech. Here, if any part of the stroke phase of the Path gesture overlapped with the relevant path speech, we considered it as synchronous. Likewise, the proportion of manner gestures that overlapped with manner speech (i.e., manner verbs, manner elaborations) were calculated.

The percentage of Path gestures that co-occurred with path speech was 82% for Turkish and 75% for English. For Manner gestures, the co-occurrence rates with manner speech were 86% for Turkish and 65% for English. That is, for both languages, a majority of Path and Manner gestures overlapped with immediately relevant speech. Chi square tests conducted on the number of manner gestures (X^2 = 2,00 df = 1, p = 0.15) and path gestures (X^2 = 2,00 df = 1, p = 0.16) that overlapped with relevant speech did not reveal significant differences between the languages in this respect. That is, in both languages, gestures overlapped with the speech that they semantically coordinate with to a similar extent.

Conclusion and Discussion

The aim of the present study was to investigate whether and how the linguistic framing of events influences gestural representations of the same events during online speaking. We attempted to go beyond the few studies that have investigated this question by providing data from a large sample, comparing two typologically different languages, English and Turkish, which use different linguistic framing for motion events (i.e., satellite-framed versus verb-framed). We also conducted tight semantic as well as temporal analyses for a large sample of gesture and speech pairs depicting various motion events from speakers of the two languages.

The results showed that in cases where there are differences in the semantic and syntactic encoding of motion event elements, gestural representations vary in ways that fit the language specific encoding differences. This was illustrated by the differences in gesture types between Turkish and English speakers' gestures that overlap with Tight versus Semi-tight speech. Specifically when speakers used one verbal clause they preferred to use one gesture that expressed both elements and when they used two verbal clauses, they were more likely to use separate gestures for manner and path. These results support previous findings (Kita & Özyürek, 2003; Özyürek & Kita, 1999), extending their generalizability to various motion event types and also providing further evidence from temporal synchrony analysis.

However, in cases where there were no language specific differences, that is, in terms of the use of Manner-only and the Path-only sentences, the gesture patterns of the speakers of the two languages looked alike. Here gestural information was found to fit the semantic encoding of the event rather than compensate or convey meaning not expressed by speech. This pattern was found to be the same in the two languages. Note that these results are unlike found for Spanish in the McNeill and Duncan (2000) study where manner gestures were found to compensate for the omission of manner content in speech. It would be revealing if a similar large scale study were conducted between Spanish and English to investigate the information coordination between speech and gesture.

The differences in the distributions of gesture types in different sentence contexts (i.e., Figures 2, 3, and 4) reveal that the differences in gesture types across languages found both in the current study as well as in Kita & Özyürek (2003) can not be explained merely due to cultural factors. Namely, it is not the case that Turkish speakers have a general preference for Manner and Path gestures but not for Conflated gestures compared to English speakers, regardless of the content of the concurrent speech. Rather, they show that gestural differences between English and Turkish speakers (Figure 4) could be directly attributed to the online choice of different semantic and syntactic encoding patterns, since gestural differences wash out when both Turkish and English speakers choose to express either only path or only manner in their speech.

Finally, the temporal synchrony analysis is in line with the rest of the findings in the sense that the information coordination between the two modalities is also reflected in the temporal synchrony of semantic information in the two channels regardless of the language spoken (McNeill, 1992). However, it is important to note that in both languages, non-typical alignments were also observed in around 25% of the cases. Further research is necessary to investigate the nature of such combinations, what they reveal, and whether they show different distributions across languages.

Further analysis is also needed to investigate how the discourse level encoding of information interacts with the analyses we have provided here. It is possible that the gestural coordination of information might be further sensitive to whether the manner and/or path information was new or old in discourse as shown in the McNeill and Duncan's (2000) analysis.

Overall, the findings of this study show that even though there are differences in the way gestures encode same events across languages, this can be explained by one and similar process underlying speech and gesture production in speakers of different languages. That is, during online speaking gestural and linguistic representations interact in such a way that gestures reflect the underlying online conceptualization that fits appropriate semantic and syntactic encoding of events. In this paper we attempted to unpack the nature of this multi-modal semantic information coordination at the sentence level and found it to be similar across speakers of different languages, at least for the two typologically different languages we have studied at length.

Acknowledgments

This research was supported by grant BCS-0002117 from the National Science Foundation to the first three authors, and by a Turkish Academy of Sciences (TUBA) Young Scientist Award to Özyürek. Substantial logistical support was provided by the Max Planck Institute for Psycholinguistics, Nijmegen, Netherlands, and by Koç University, Istanbul, Turkey. Audiences at the Max Planck Institute for Psycholinguistics offered helpful comments and discussion on previous versions of this paper. Tilbe Göksun helped in establishing reliability.

Note

1. This is not to say that gestures will not depict any representations not expressed in speech. The point here is that their representations will be influenced by the linguistic encoding possibilities in speech.

References

Brugman, Hennie & Sotaro Kita (1995). Impact of digital video technology on transcription: a case of spontaneous gesture transcription. *KODIKAS/CODE: Ars Semeiotica, An international journal of semiotics, 18,* 95–112.

Clark, Herbert (1996). *Using language.* Cambridge: Cambridge University Press.

Garrett, Merill (1982). Production of speech: Observations from normal and pathological language use. In Willem Ellis (Ed.), *Normality and pathology in cognitive functions* (pp. 19–76). London: Academic Press.

Goldin-Meadow, Susan (2004). *Hearing gesture: How our hands help us think.* Cambridge: Harvard University Press.

Kendon, Adam (1980). Gesticulation and speech: Two aspects of the process of utterance. In Mary R. Key (Ed.), *The relation between verbal and nonverbal communication* (pp. 207–227). The Hague: Mouton.

Kendon, Adam (2004). *Gesture. Visible action as utterance.* Cambridge: Cambridge University Press.

Kita, Sotaro (2000). How representational gestures help speaking. In David McNeill (Ed.), *Language and gesture* (pp. 162–185). Cambridge: Cambridge University Press.

Kita, Sotaro, Ingeborg van Gijn, & Harry van der Hulst (1998). Movement Phases in signs and co-speech gestures, and their transcription by human coders. In Ipke Wachsmuth & Martin Fröhlich (Eds.), *Gesture and sign language in human–computer interaction, International Gesture Workshop Bielefeld, Germany, September 17–19, 1997, Proceedings.* Lecture Notes in Artificial Intelligence, Volume 1371 (pp. 23–35). Berlin: Springer-Verlag.

Kita, Sotaro & Asli Özyürek (2003). What does cross-linguistic variation in semantic coordination of speech and gesture reveal? Evidence for an interface representation of spatial thinking and speaking. *Journal of Memory and Language, 48,* 16–32.

Lemmens, Maarten (2002). The semantic network of Dutch posture verbs. In John Newman (Ed.), *The linguistics of sitting, standing, and lying* (pp. 103–139). Amsterdam: Benjamins.

Levinson, Stephen & Sergio Meira (2003). 'Natural concepts' in the spatial topological domain — adpositional meanings in cross-linguistic perspective: An exercise in semantic typology. *Language,* 79(3), 485–516.

Levelt, Pim (1989). *Speaking.* Cambridge: MIT Press.

McNeill, David (1985). So you think gestures are nonverbal? *Psychological Review, 92,* 350–371.

McNeill, David (1992). *Hand and mind.* Chicago: University of Chicago Press.

McNeill, David (2000). Analogic/analytic representations and cross-linguistic differences in thinking for speaking. *Cognitive Linguistics,* 11 (1/2), 43–60.

McNeill, David & Susan Duncan (2000). Growth points in thinking-for-speaking. In David McNeill (Ed.), *Language and gesture* (pp. 141–161). Cambridge: Cambridge University Press.

Müller, Cornelia (1998). *Redebegleitende Gesten: Kulturgeschichte-Theorie-Spachvergleich.* Berlin: Berlin Verlag Arno Spitz.

Özyürek, Asli & Sotaro Kita (1999). Expressing manner and path in English and Turkish: Differences in speech, gesture, and conceptualization. In Martin Hahn & Scott Stoness (Eds.), *Proceedings of the twenty first annual conference of the Cognitive Science Society* (pp. 507–512). Mahwah, NJ: Lawrence Erlbaum.

Özyürek, Asli, Sotaro Kita, & Shanley Allen (2001). *Tomato man movies: Stimulus kit designed to elicit manner, path and causal constructions in motion events with regard to speech and gestures*. Nijmegen, The Netherlands: Max Planck Institute for Psycholingusitics, Language and Cognition group.

Slobin, Dan (1987). Thinking for speaking. In Jon Aske, Natasha Beery, Laura Michaelis, & Hana Filip (Eds.), *Proceedings of the 13th annual meeting of the Berkeley Linguistic Society* (pp. 435–445). Berkeley: Berkeley Linguistic Society.

Slobin, Dan (1996). From "thought and language" to "thinking for speaking". In John Gumperz & Stephen Levinson (Eds.), *Rethinking linguistic relativity* (pp. 70–96). Cambridge: Cambridge University Press.

Talmy, Len (1985). Semantics and syntax of motion. In Timothy Shopen (Ed.), *Language typology and syntactic description, Vol.3, Grammatical categories and the lexicon* (pp. 57–149). Cambridge: Cambridge University Press.

The two faces of gesture

Language and thought

Susan Goldin-Meadow
University of Chicago

Gesture is typically produced with speech, forming a fully integrated system with that speech. However, under unusual circumstances, gesture can be produced completely on its own — without speech. In these instances, gesture takes over the full burden of communication usually shared by the two modalities. What happens to gesture in these two very different contexts? One possibility is that there are no differences in the forms gesture takes in these two contexts — that gesture is gesture no matter what its function. But, in fact, that's not what we find. When gesture is produced on its own, it assumes the full burden of communication and takes on a language-like form, with sentence-level ordering rules, word-level paradigms, and grammatical categories. In contrast, when gesture is produced in conjunction with speech, it shares the burden of communication with speech and takes on a global imagistic form, often conveying information not found anywhere in speech. Gesture thus changes its form according to its function.

When people talk, they gesture. Indeed, it is almost impossible for people to talk naturally without gesturing. When gesture is produced along with speech, it forms an integrated system with that speech, sharing with it the burden of communication. However, there are other situations in which gesture is produced on its own. In these contexts, gesture assumes the full burden of communication. For example, congenitally deaf children whose profound hearing losses prevent them from acquiring the spoken language that surrounds them cannot use speech to communicate. If, in addition, these children are not exposed to a conventional sign language, they are also unable to use conventional sign to communicate. Spontaneously created gesture is the only accessible means of communicating that these children have, and they use it.

What form does gesture assume in these two very different contexts? One might guess that gesture would assume the same form when produced with speech and without it. But this guess would be wrong. In fact, gesture looks quite different when it shares the burden of communication with speech, compared to when it assumes the full burden of communication on its own. When produced along with speech, gesture is framed by that speech. It takes on a global and holistic form that is interpretable only within the framing that speech provides. In contrast, when produced on its own, gesture assumes the discrete and segmented form characteristic of all linguistic systems. It becomes language-like. Thus, gesture changes its form as it changes its context and its function.

Gesture without speech

Background on deafness and language-learning

When deaf children are exposed to sign language from birth, they learn that language as naturally as hearing children learn spoken language (Newport & Meier, 1985). However, 90% of deaf children are not born to deaf parents who could provide early access to sign language. Rather, they are born to hearing parents who, quite naturally, expose their children to speech. Unfortunately, it is extremely uncommon for deaf children with severe to profound hearing losses to acquire spoken language without intensive and specialized instruction. Even with instruction, their acquisition of speech is markedly delayed (Conrad, 1977; Mayberry, 1992).

We have studied ten children whose severe hearing losses prevented them from acquiring spoken language naturally. Moreover, their parents had decided to educate them in oral schools where sign systems are neither taught nor encouraged (Goldin-Meadow, 2003a). At the time of our observations, the children ranged in age from 1;2 to 4;10 (years;months) and had made little progress in oral language, occasionally producing single recognizable words but never combining those words into sentences. In addition, they had not been exposed to a conventional sign system of any sort (e.g., American Sign Language or a manual code of English). The children thus knew neither sign nor speech.

Sentence level structure: syntax

All of the children used gesture without speech to communicate with the hearing individuals in their worlds. Moreover, all of the children combined their gestures into strings. For example, one child combined a point at a grape with an "eat" gesture to comment on the fact that grapes can be eaten, and then later combined the "eat" gesture with a point at the experimenter to invite her to lunch.

Moreover, the gesture strings that the deaf children produced functioned in a number of respects like the sentences of early child language. On this basis, these strings warrant the label "sentence." The children produced gesture sentences characterized by two types of *surface regularities:* (1) regularities in the production and deletion of elements in a sentence, and (2) regularities in the position within the sentence that those elements occupied.

As an example of the first type of surface regularity, although the children rarely produced gestures for all of the possible thematic roles that could be conveyed within a proposition, they were not haphazard in their selection of which roles to convey in gesture and which to omit. For example, the children were equally likely to produce gestures for the *intransitive actor* (e.g., the mouse in a sentence describing a mouse running to his hole) as for the *patient* (e.g., the cheese in a sentence describing a mouse eating cheese), and were far more likely to produce either of these than gestures for the *transitive actor* (e.g., the mouse in a sentence describing a mouse eating cheese) (Goldin-Meadow & Mylander, 1984). In this way, the likelihood of production served to systematically distinguish among thematic roles and thus mark those roles, an important function of grammatical devices. It is also worth noting that the particular pattern found in the deaf children's gestures — patients and intransitive actors marked in the same way and both different from transitive actors — is an analogue of a structural case-marking pattern found in naturally occurring human languages, ergative languages (cf. Dixon, 1979; Silverstein, 1976).

As an example of the second type of surface regularity, the children distinguished among the thematic roles they did express by placing the gesture for a given role in a particular position in a gesture sentence; that is, the gestures the children produced within their sentences were not produced in haphazard sequence but rather appeared to follow a small set of gesture order regularities (Goldin-Meadow & Mylander, 1984). For example, in a sentence commenting on the child's intention to throw a toy grape, the child first produced a gesture for the grape, the *patient* (typically a pointing gesture at the grape but, at times, an iconic gesture for the grape) before producing a gesture for the *act* "throw" (an iconic gesture). In general, the gesture for the object playing a patient role tended to *precede* the gesture for the act. As a second example of a gesture order, gestures for an object playing the role of recipient or goal tended to *follow* gestures for the act; e.g., in a sentence used to request that an object be moved to a puzzle, the child produced a gesture for the *act* "transfer" (an iconic gesture) before producing a gesture for the *recipient*, "puzzle" (a pointing gesture).

In addition to these regularities at the surface level, the children's gesture sentences were organized at *underlying levels.* Each sentence expressed one or more frames composed of a predicate and 1, 2, or 3 arguments (Goldin-Meadow, 1985: 215–219; Feldman, Goldin-Meadow, & Gleitman, 1978: 385–388). For example, all of the children produced "transfer" or "give" predicates with an inferred frame containing 3 arguments — the actor, patient, and recipient (e.g., *you/sister* give *duck* to *her/Susan*). The children also produced two types of 2-argument predicates: transitive predicates such as "drink" with a frame containing the actor and patient (e.g., *you/Susan* drink *coffee*), and intransitive predicates such as "go" with a frame containing the actor and recipient (e.g., *I/child* go *downstairs*). Finally, the children produced predicates such as "sleep" or "dance" with a 1-argument frame containing only the actor (e.g., *he/father* sleep).

The children frequently concatenated more than one predicate frame within the bounds of a single sentence — that is, they produced complex as opposed to simple sentences, thus demonstrating the important property of recursion in their gesture systems (e.g., a "climb" gesture, followed by a "sleep" gesture, followed by a point at a horse, to comment on the fact that the horse in a picture climbed up the house and then slept). Recursion gives language its generative capacity and is found in all natural language systems. Importantly, when the children concatenated more than one predicate frame within a single sentence, they did so systematically, allocating one "slot" in underlying structure to the arguments and predicates playing a role in both frames (e.g., 'he/horse' is assigned only one slot in underlying structure in the above sentence in which the horse played the actor role in both of the predicates of the concatenated frames, '*he/horse* climbs and sleeps'; Goldin-Meadow, 1982).

Thus, the deaf children conjoined the gestures they produced into sentences characterized by surface regularties (regularities in likelihood of production and deletion and in gesture order), as well as regularities at underlying levels (predicate frames underlying each simple and complex gesture sentence). The gesture strings could therefore be said to conform to a syntax, albeit a simple one.

Word level structure: morphology

The deaf children's gestures not only formed parts of longer sentence-units but they themselves were made up of smaller parts. For example, to request the experimenter to lay a penny down flat on a toy, one deaf child produced a downward motion with his hand shaped like an O. In itself this could be a global gesture presenting the shape and trajectory as an unanalyzed whole. The experimenter pretended not to understand and, after several repetitions, the child factored the gesture into its components: first he statically held up the gesture for a round object (the O handshape) and then, quite deliberately and with his hand no longer in the O shape but exhibiting a flat palm, made the trajectory for downward movement. The original gesture was thus decomposed into two elements. This example hints at the presence of a system of linguistic segments in which the complex meaning of "round-thing-moving-downward" is broken into components and the components combined into a gesture. Although the experimenter's feigned lack of understanding was undoubtedly important in getting the child to decompose his gesture at that particular moment, the important point is that when the child did break his gesture into parts, those parts were elements of a wider system — one that accounted for virtually all of the gestures that this child produced.

The child had thus devised a morphological system in which each gesture was a complex of simpler elements (Goldin-Meadow & Mylander, 1990; see also Singleton, Morford, & Goldin-Meadow, 1993). As an example of how this child's gestures formed a system of contrasts, a *CMedium* handshape (the hand shaped in a C with the fingers 1–3 inches from the thumb) meant 'handle an object 2–3 inches wide,' and a *Revolve* motion meant 'rotate around an axis'. When combined, these two components created a gesture whose meaning was a composite of the two meanings — 'rotate an object 2–3 inches wide' (e.g., twist a jar lid). When the same *CMedium* handshape was combined with a different motion, a *Short Arc* (meaning 'reposition'), the resulting combination had a predictably different meaning — 'change the position of an object 2–3 inches wide' (e.g., tilt a cup). As a result, the child's gestures can be said to conform to a framework or system of contrasts.

We have analyzed the gesture systems of four deaf children at this level (Goldin-Meadow, Mylander, & Butcher, 1995), and found that all four produced gestures that could be characterized by paradigms of handshape and motion combinations. Thus, each child:

- used a limited set of discrete handshape and motion forms, that is, the forms were categorical rather than continuous;
- consistently associated each handshape or motion form with a particular meaning (or set of meanings) throughout the corpus, that is, each form was meaningful;
- produced most of the handshapes with more than one motion, and most of the motions with more than one handshape, that is, each handshape and motion was an independent and meaningful morpheme that could combine with other morphemes in the system to create larger meaningful units — the system was combinatorial.

Although similar in many respects, the gesture systems produced by these four children were sufficiently different to suggest that the children had introduced relatively arbitrary — albeit still iconic — distinctions into their systems. For example, in contrast to the first child and one other who used the *CMedium* handshape to represent objects 2–3 inches in width (e.g., a cup or a box), the two other children used the same *CMedium* handshape to represent objects that were slightly smaller, 1–2 inches in width (e.g., a banana or a toy soldier, Goldin-Meadow et al., 1995). The fact that there were differences in the ways the children defined a particular morpheme suggests that there were choices to be made (although all of the choices still were transparent with respect to their referents). Moreover, the choices that a given child made could not be determined without knowing that child's individual system. In other words, we cannot predict the precise boundaries of a child's morphemes without knowing that child's individual system. In this sense the deaf children's gesture systems can be said to be *arbitrary*.

Grammatical categories

Do the deaf children have the category 'noun' and, if so, does it contrast grammatically with the categories 'verb' and 'adjective'? To address this question, we examined all of the iconic gestures that one deaf child produced over a two-year period (Goldin-Meadow, Butcher, Mylander, & Dodge, 1994). We identified iconic gestures used to focus attention on the discourse topic as *nouns*, and iconic gestures used to comment on that topic as predicates — *verbs* if the particular comment described an action, *adjectives* if it described an attribute. We found that gestures playing noun-like roles were distinguished from those playing verb-like roles in two ways — by the form of the gesture (akin to a morphological marking), and by its position in a gesture sentence (akin to a syntactic marking). The distinction between nouns and verbs is most strikingly seen in gestures used in both roles. For example, if the child used a "twist" gesture to focus attention on a jar as the discourse topic (i.e., as a noun), the gesture was likely to be abbreviated in form (one twist of the hand rather than several — an alteration internal to the gesture and therefore a morphological marking) and not inflected, and was likely to *precede* a deictic pointing gesture at the jar (a relation across gestures and therefore a syntactic marking). If, on another occasion, that same stem "twist" was used to say something about the jar (i.e., as a verb), the gesture was likely to be inflected in form (produced in a space near the jar, the patient of this particular predicate, rather than in neutral space — a morphological marking) and not abbreviated, and was likely to *follow* a deictic pointing gesture at the jar (a syntactic marking).

While all languages distinguish nouns and verbs, only certain languages make a further distinction between nouns and verbs and a third class, the class of adjectives (Schachter, 1985). We found that, in the deaf child's system, gestures used as adjectives were treated like nouns with respect to morphology (i.e., adjectives tended to be abbreviated but not inflected), but like verbs with respect to syntax (i.e., adjectives tended to follow pointing gestures rather than precede them). The deaf child's adjective gestures consequently appear to behave as adjectives do in natural languages — sharing some morpho-syntactic properties with nouns and others with verbs (cf. Thompson, 1988). Maintaining a distinction between nouns, verbs, and adjectives thus appears to be a property of gestures when they assume the full burden of communication.

Where does this language-like structure come from?

The deaf children are inventing their gesture systems without input from a conventional language model. They are not, however, inventing their gesture systems in a vacuum. Like all speakers (Goldin-Meadow, 2003b), the children's hearing parents gesture when they talk, and the deaf children have access to those gestures. The children could be modeling their gesture systems after the gestures that their parents produce. Although perfectly reasonable, this hypothesis is incorrect. When we analyze the hearing parents' gestures with the same tools that we use to analyze the deaf children's gestures, we find that the two sets of gestures have little in common.

Beginning with sentence level structure, we analyzed the gestures that the hearing mothers of six of our deaf children produced as they talked to their children, looking for production probability and gesture order patterns (Goldin-Meadow & Mylander, 1983, 1998). We found that the mothers rarely combined their gestures with other gestures and thus rarely produced gesture "sentences". Moreover, the few gesture sentences that they did produce patterned differently, in terms of both production probability and gesture order, from their children's gestures. Unlike the deaf children, all of whom displayed the same pattern across both devices (production probability and gesture order), the mothers showed no uniformity, either across individuals or across devices within an individual. In addition, the mothers began using recursion in their gesture sentences *after* their children and used it significantly less often than their children.

To examine word level structure, we analyzed the gestures that the hearing mothers of four of our deaf children produced (Goldin-Meadow et al., 1995) and found that, here again, the mothers' gestures were quite distinct from their children's. Each mother used her gestures in a more restricted way than her child, omitting many of the morphemes that the child produced (or using the ones she did produce more narrowly than the child), and omitting completely many of the handshape/motion combinations that the child produced. In addition, while there was good evidence that the gestures of each deaf child could be characterized in terms of handshape and motion components which mapped onto a variety of related objects and a variety of related actions, respectively, there was no evidence that the mothers ever went beyond mapping gestures as wholes onto entire events — that is, the mother's gestures did not appear to be organized in relation to one another to form a system of contrasts. Finally, when the mothers' gestures were analyzed with the same procedures used to analyze the children's gestures (that is, when the mother's gestures were treated as a system unto itself), the resulting systems for each mother did not capture her child's gestures well at all. Most importantly, the arbitrary differences that were found across the children's systems could *not* be traced to the mothers' gestures, but seemed instead to be shaped by the early gestures that the children themselves created. In other words, the differences could be traced to the gestural input that the children provided for themselves rather than to gestural input that their mothers provided for them.

Finally, we found that the mother of the deaf child who distinguished among nouns, verbs, and adjectives did not use the same morphological and syntactic devices in her gestures that her child used in his to make these distinctions (Goldin-Meadow et al., 1994). Indeed, certain of the devices that the child used to distinguish these categories (abbreviation and gesture order) either were not used at all or were not used distinctively by mother. These devices were therefore likely to have been initiated by the child. The child's third device (inflection) was used by mother; however, the child's inflections patterned systematically with the predicate structure of the verb and consistently marked entities playing particular thematic roles in those predicates — that is, they functioned as part of a system — while mother's did not. Thus, while the child may have used the gestural input his mother provided as a starting point for part of his system, he went well beyond that input — fashioning it into an integral component of the system and grammaticizing it as he did so.

Two points are worth noting. First, the mothers' gestures could not have served as a model for the structure found in their children's gestures. Second, the deaf children's gestures were not forced by the modality to assume sentence and word level structure — the mothers' gestures were also produced in the manual modality yet they did *not* assume language-like forms at these levels.

I suggest that the mothers' gestures appear random and without structure only because we have examined them through the wrong lens. These gestures were produced along with speech and were meant to be interpreted in the context of speech. Constrained by the speech with which they co-occurred, the mothers' gestures were not at liberty to assume the language-like form that characterized their children's gestures (cf. Goldin-Meadow, McNeill, & Singleton, 1996). They assumed instead the form that all speech-occurring gestures take on. We look next at the gesture that accompanies speech in the way it was meant to be seen — with speech.

Gesture with speech

Gesture and speech form an integrated system

Gesture is pervasive. It occurs with speech in all contexts and, importantly, it is not just hand-waving. Unlike the deaf children's gestures which resemble beads on a string, the gestures that hearing speakers produce along with their talk are global and synthetic in form (McNeill, 1992). For example, a deaf child might point at a jar and then, with a C-shaped hand, produce a twisting motion several times in the direction of the jar to comment on jar-opening. A hearing speaker, by contrast, would be more likely to loosely rotate a floppy hand several times in front of the body while saying the word "open". Nevertheless, and despite their less well-articulated handshapes and motions, the gestures that accompany speech do convey substantive meaning (Clark, 1996; Goldin-Meadow et al., 1996; Kendon, 1980; McNeill, 1992).

Gestures are integrated both semantically and temporally with the speech they accompany (McNeill, 1992). In the next sections, I focus on one compelling aspect of the gesture-speech relationship — the fact that gesture reflects the cognitive state of the speaker (Goldin-Meadow, 2003b). I argue that gesture can provide a unique perspective on that state, one that is not reflected in speech.

Gesture offers a unique perspective on the cognitive state of the speaker

I begin with an example. Consider a child asked to justify his responses to a Piagetian conservation task. The child is shown two rows, each containing six checkers and is asked to verify that the rows have the same number of checkers. The experimenter then spreads the checkers in one of the rows out, and again asks whether the rows have the same or a different number of checkers. The child says "different." To justify his belief that the number of checkers has changed, the child indicates in speech that "you moved them." However, he does *not* refer to moving the checkers in gesture. Rather, he uses a pointing gesture to pair the first checker in one row with the first checker in the other, the second with the second, and so on. In other words, he indicates the one-to-one correspondence between the checkers in the two rows *in gesture*, while at the same time describing how the experimenter moved the checkers *in speech*. The child has produced a gesture-speech mismatch — he has conveyed information in gesture that is different from the information he conveyed in speech.

Gesture-speech mismatch is a widespread phenomenon. It occurs in many cognitive tasks and over a large age range: in toddlers experiencing vocabulary spurts (Gershkoff-Stowe & Smith, 1997); preschoolers explaining games (Evans

& Rubin, 1979); elementary school children explaining mathematical equations (Perry, Church, & Goldin-Meadow, 1988) and seasonal change (Crowder & Newman, 1993); children and adults discussing moral dilemmas (Church et al., 1995); adolescents explaining Piagetian bending-rods tasks (Stone, Webb, & Mahootian, 1991); and adults explaining gears (Perry & Elder, 1996; Schwartz & Black, 1996) and problems involving constant change (Alibali et al., 1999).

In addition to being pervasive, mismatches can be uniquely informative. We examined the problem-solving strategies fourth grade children gave when explaining their solutions to math problems of the following type, 4+5+3=__+3 (Goldin-Meadow, Alibali, & Church, 1993). If a child produced a strategy for solving the problem *only* in gesture across all six problems, that strategy was assigned to the "gesture only" repertoire. If, however, the child also produced that strategy in speech at some point over the six problems, it was assigned to the "speech+gesture" repertoire. We followed the same criteria in assigning strategies to the "speech only" repertoire.

The children varied in the number of gesture-speech mismatches they produced on the math task. Moreover, the children who produced many mismatches on the task had much larger gesture-only repertoires than those who produced few. But the two groups did not differ in their speech+gesture or speech-only repertoires. What this means is that mismatchers had a larger number of different strategies for solving the task at their disposal than matchers, and that all of the "extra" strategies could be found *only* in gesture. If we want to know what mismatchers understand about a task, we cannot just listen to them — we have to look at them too.

Thus, gesture can convey information that is not found *anywhere* in the speaker's verbal repertoire. In addition, as I show in the next three sections, the relation between gesture and speech can provide insight into how speakers learn, solve problems and remember.

Gesture can predict who will learn

To explore whether the relation between gesture and speech predicts who will profit from instruction, we took children who failed initially on either a math task (Perry et al., 1988) or a conservation task (Church & Goldin-Meadow, 1986). We first asked the children how they solved the problems on the task and, on the basis of their explanations, classified the children into those who produced many mismatches and those who produced few. We then gave all of the children instruction in how to solve the math or conservation task, followed by yet another test of their knowledge. We found that, on both the math and conservation tasks, children who produced many mismatches during their initial explanations were far more likely to show significant gains on the follow-up test after instruction than children who produced few mismatches (see also Pine, Lufkin, & Messer, 2004, who replicated the phenomenon on a balancing task).

Why might children who produce many mismatches be more ready to learn than children who produce few? As described in the previous section, children who mismatch often on a task have more substantive knowledge about that task than children who mismatch less often. However, all of this additional knowledge is accessible only to gesture and not to speech. Thus, the extra knowledge cannot be explicitly articulated and cannot be integrated into the child's framework for solving the problem. Mismatchers have the pieces in place to make progress on a task, but have not yet pulled those pieces together. Instruction provides the impetus and perhaps the framework for reorganizing the pieces, and leads to success on the task. The mismatch between gesture and speech allows us to tell, before the fact, who will profit from instruction and who will not.

Gesture can predict how a problem will be solved

To explore whether an adult's gestures predict how that adult will solve a problem, we gave adults a series of word problems of the following sort: "A bookcase has six shelves; the number of books on each successive shelf increases by a constant number. If there are 15 books on the top shelf and 45 on the bottom, how many books total are there?" This problem can be solved in one of two ways — in terms of discrete units of books added, or in terms of a continuous rate of books added. Discrete verbal descriptions are typically accompanied by short, choppy, step-like gestures, i.e., iterations of discrete movements. Continuous verbal descriptions are typically accompanied by longer, more flowing gestures, i.e., smooth curving movements. We ask the adults, first, to restate the problem for us and, then, to describe how they would go about solving the problem. We then attempted to predict how speakers would solve the problem as a function of the gestures and speech that they produced in their initial problem descriptions (Alibali et al., 1999).

When we looked at the initial descriptions adults gave of the problems, we found that they often produced gestures that reinforced their verbal descriptions, i.e., they produced discrete gestures along with discrete verbal descriptions, and continuous gestures along with continuous verbal descriptions. However, at times, the speakers' gestures either were neutral with respect to the verbal description they accompanied (neither discrete nor continuous), or they conflicted with the verbal description (discrete gestures with a continuous verbal description, or continuous gestures with a discrete verbal description).

Interestingly, the strategy speakers said they would use to solve the problem was much more likely to match the verbal strategy they used in their initial description of that problem when their gestures also conveyed this same strategy. For example, speakers were significantly more likely to say that they would solve the problem using a discrete strategy if their speech and gestures in their initial problem descriptions were both discrete than if their gestures were continuous (or neutral) and their speech was discrete. It is very likely that the adults were completely unaware of the conflicting information that they displayed in gesture in their initial description of the problem. Nonetheless, these unacknowledged difficult-to-integrate pieces of information appeared to have an impact on the adults' plans for solving the problem. And, once again, we would not have had access to these plans had we not looked at the speakers while we listened to them.

Gesture can lighten cognitive load and thus improve memory

To explore whether gesturing can affect how much a speaker will remember, we asked adults and children to remember a list of words (for the children) or letters (for the adults) while explaining their answers to a math problem (Goldin-Meadow, Nusbaum, Kelly, & Wagner, 2001). The children and adults did their explaining under two conditions — on half the problems, they were allowed to move their hands freely; on the other half, they were asked to keep their hands still on the table (gesture *per se* was not mentioned at any time during the study).

A priori we might expect gesturing to add to a speaker's cognitive load. After all, a speaker who is producing gestures while talking must coordinate the two modalities; doing two things at once, in principle, ought to take more cognitive effort than doing only one. If so, speakers should remember *fewer* words when they gestured on their explanations than when they did not gesture. If, however, gesture and speech form a synergistic system in which effort expended in one modality reduces effort expended overall, speakers should remember *more* words when they gestured on their explanations that when they did not gesture.

We found that both children and adults remembered more words when they gestured than when they did not gesture, suggesting that gesturing can actually lighten a speaker's cognitive load. It is possible, however, that rather than gesturing lightening the load, being told to keep one's hands still is adding to the load. Our data provided us with a simple way of addressing this concern. Some of the adults and children did not gesture on all of the problems that they explained when their hands were free. Thus, for these speakers, we effectively had three conditions: gesture by choice, no gesture by choice, and no gesture by instruction. If the instructions themselves are creating a cognitive load for the speakers, speakers should remember significantly fewer words when told not to gesture than when they spontaneously chose not to gesture. However, this is not the pattern we found. Adults and children remembered the same number of words whether they were told not to gesture or chose by themselves not to gesture, and this number was significantly smaller than the number of words they remembered when they chose to gesture.

Perhaps counter intuitively, gesturing appears to *save* cognitive resources, resources that can then be allocated to another task (e.g., a memory task). Thus, gesturing may not only reflect a speaker's cognitive state but, in reducing cognitive load, it may also play a role in shaping that state.

The two faces of gesture

Gesture is chameleon-like in its form. Moreover, the form gesture assumes appears to be tied to the function it serves. When gesture assumes the full burden of communication, acting on its own without speech, it takes on a language-like form. It has sentence-level structure, word-level paradigms, and grammatical categories — all forms that are *not* found when gesture is produced along with speech.

When gesture shares the burden of communication with speech, it loses its language-like structure, assuming instead a global and synthetic form. Although not language-like in structure when it accompanies speech, gesture stills forms an important part of language. It conveys information imagistically and, as such, has access to different information than does the verbal system. Gesture thus allows speakers to convey thoughts that may not easily fit into the categorical system that their conventional language offers (Goldin-Meadow & McNeill, 1999). Gesture therefore offers us a window into the mind that is distinct from the window that speech offers. Indeed, it is only by looking at *both* gesture and speech that we can predict how people learn, remember, and solve problems. Although not language-like in form, gesture is nevertheless an integral part of language, cropping up whenever there is talk. As language researchers, we cannot afford to ignore it.

Acknowledgement

This research was supported by grants from the National Science Foundation (BNS 8810879), the National Institute of Deafness and Other Communication Disorders (R01 DC00491), the National Institutes of Child Health and Human Development (R01 HD47450), and the Spencer Foundation.

References

Alibali, Martha W., Miriam Bassok, Karen O. Solomon, Sharon E. Syc, & Susan Goldin-Meadow (1999). Illuminating mental representations through speech and gesture. *Psychological Science, 10*, 327–333.

Church, Ruth Breckinridge & Susan Goldin-Meadow (1986). The mismatch between gesture and speech as an index of transitional knowledge. *Cognition, 23*, 43–71.

Church, Ruth Breckinridge, Kimberly Schonert-Reichl, Nancy Goodman, Spencer D. Kelly, & Saba Ayman-Nolley (1995). The role of gesture and speech communication as reflections of cognitive understanding. *Journal of Contemporary Legal Issues, 6*, 123–154.

Clark, Herbert H. (1996). *Using language*. New York: Cambridge University Press.

Conrad, R. (1977). Lip-reading by deaf and hearing children. *British Journal of Educational Psychology, 47*, 60–65.

Crowder, Elaine & Denis Newman (1993). Telling what they know: The role of gesture and language in children's science explanations. *Pragmatics and Cognition, 1*, 341–376.

Dixon, Robert M. W. (1979). Ergativity. *Language, 55*, 59–138.

Evans, Mary A. & Kenneth H. Rubin (1979). Hand gestures as a communicative mode in school-aged children. *The Journal of Genetic Psychology, 135*, 189–196.

Feldman, Heidi, Susan Goldin-Meadow, & Lila R. Gleitman (1978). Beyond Herodotus: The creation of language by linguistically deprived deaf children. In A. Lock (Ed.), *Action, symbol, and gesture: The emergence of language*. New York: Academic Press.

Gershkoff-Stowe, Lisa & Linda B. Smith (1997). A curvilinear trend in naming errors as a function of early vocabulary growth. *Cognitive Psychology, 34*, 37–71.

Goldin-Meadow, Susan (1982). The resilience of recursion: A study of a communication system developed without a conventional language model. In E. Wanner & L. R. Gleitman (Eds.), *Language acquisition: The state of the art* (pp. 51–77). New York: Cambridge University Press.

Goldin-Meadow, Susan (1985). Language development under atypical learning conditions: Replication and implications of a study of deaf children of hearing parents. In K. Nelson (Ed.), *Children's language* (Vol. 5; pp. 197–245). Hillsdale, N.J.: Lawrence Erlbaum & Associates.

Goldin-Meadow, Susan (2003a). *Resilience of language*. New York: Psychology Press.

Goldin-Meadow, Susan (2003b). *Hearing gesture*. Cambridge, MA: Harvard University Press.

Goldin-Meadow, Susan, Martha W. Alibali, & Ruth Breckinridge Church (1993). Transitions in concept acquisition: Using the hand to read the mind. *Psychological Review, 100*, 279–297.

Goldin-Meadow, Susan, Cynthia Butcher, Carolyn Mylander, & Mark Dodge (1994). Nouns and verbs in a self-styled gesture system: What's in a name? *Cognitive Psychology, 27*, 259–319.

Goldin-Meadow, Susan & David McNeill (1999). The role of gesture and mimetic representation in making language the province of speech. In Michael C. Corballis & Stephen Lea (Eds.), *The descent of mind* (pp. 155–172). Oxford: Oxford University Press.

Goldin-Meadow, Susan, David McNeill, & Jenny Singleton (1996). Silence is liberating: Removing the handcuffs on grammatical expression in the manual modality. *Psychological Review, 103*, 34–55.

Goldin-Meadow, Susan & Carolyn Mylander (1983). Gestural communication in deaf children: The non-effects of parental input on language development. *Science, 221*, 372–374.

Goldin-Meadow, Susan & Carolyn Mylander (1984). Gestural communication in deaf children: The effects and non-effects of parental input on early language development. *Monographs of the Society for Research in Child Development, 49*, 1–121.

Goldin-Meadow, Susan & Carolyn Mylander (1990). The role of a language model in the development of a morphological system. *Journal of Child Language, 17*, 527–563.

Goldin-Meadow, Susan & Carolyn Mylander (1998). Spontaneous sign systems created by deaf children in two cultures. *Nature, 91*, 279–281.

Goldin-Meadow, Susan, Carolyn Mylander, & Cynthia Butcher (1995). The resilience of combinatorial structure at the word level: Morphology in self-styled gesture systems. *Cognition, 56*, 195–262.

Goldin-Meadow, Susan, Howard Nusbaum, Spencer Kelly, & Susan Wagner (2001). Explaining math: Gesturing lightens the load. *Psychological Sciences, 12*, 516–522.

Kendon, Adam (1980). Gesticulation and speech: Two aspects of the process of utterance. In Mary R. Key (Ed.), *Relationship of verbal and nonverbal communication* (pp. 207–228). The Hague: Mouton.

Mayberry, Rachel I. (1992). The cognitive development of deaf children: Recent insights. In Sidney Segalowitz & Isabelle Rapin (Eds.), *Child neuropsychology, Volume 7, Handbook of neuropsychology* (pp. 51–68), F. Boller & J. Graffman (Series eds.). Amsterdam: Elsevier.

McNeill, David (1992). *Hand and mind*. Chicago: University of Chicago Press.

Newport, Elissa L. & Richard P. Meier (1985). The acquisition of American Sign Language. In Dan I. Slobin (Ed,), *The cross-linguistic study of language acquisition, Vol. 1: The data*. Hillsdale, N.J.: Lawrence Erlbaum.

Perry, Michelle, Ruth Breckinridge Church, & Susan Goldin-Meadow (1988). Transitional knowledge in the acquisition of concepts. *Cognitive Development, 3*, 359–400.

Perry, Michelle & Anastasia D. Elder (1996). Knowledge in transition: Adults' developing understanding of a principle of physical causality. *Cognitive Development, 12*, 131–157.

Pine, Karen J., Nicola Lufkin, & David Messer (2004). More gestures than answers: Children learning about balance. *Developmental Psychology, 40*, 2059–2067.

Schachter, Paul (1985). Parts-of-speech systems. In Timothy Shopen (Ed.), *Language typology and syntactic description: Clause structure* (Vol. 1; pp. 3–61). Cambridge, MA: Cambridge University Press.

Schwartz, Dan L. & John B. Black (1996). Shuttling between depictive models and abstract rules: Induction and fallback. *Cognitive Science, 20*, 457–497.

Silverstein, Michael (1976). Hierarchy of features and ergativity. In R.M.W. Dixon (Ed.), *Grammatical categories in Australian languages* (pp. 112–171). Canberra: Australian Institute of Aboriginal Studies.

Singleton, Jenny L., Jill P. Morford, & Susan Goldin-Meadow (1993). Once is not enough: Standards of well-formedness in manual communication created over three different timespans. *Language, 69*, 683–715.

Stone, Addison, Rebecca Webb, & Shahrzad Mahootian (1991). The generality of gesture-speech mismatch as an index of transitional knowledge: Evidence from a control-of-variables task. *Cognitive Development, 6*, 301–313.

Thompson, Sandra A. (1988). A discourse approach to the cross-linguistic category 'adjective'. In John A. Hawkins (Ed.), *Explaining language universals* (pp. 167–185). Cambridge, MA: Basil Blackwell.

Part IV

Future directions

Gestures in human and nonhuman primates

Why we need a comparative view*

Cornelia Müller
European University Viadrina, Frankfurt (Oder)

The present article offers a condensed overview, of why a comparative view on gestures in human and nonhuman primates offers insights for both researchers of human as well as of nonhuman primates. It is argued that a comparative view may further contribute to the debate over the evolution of language but that in addition it may also enhance understanding of the relation of language and gesture in humans. The article sketches programmatic issues, which are summarized in the list of framing questions for the workshop and this volume on "Gestural communication in nonhuman and human primates"; it is aimed to clarify conceptual and methodological prerequisites and to offer points of departure for future comparative research.

In this article I wish to provide a brief but systematic plea for comparative studies of gesture in humans and nonhumans, and set out the prerequisits and the conceptual and methodological grounds for such a comparative view. I shall elaborate on some of the programmatic issues raised in the Introduction to this volume and offer a condensed overview of specific perspectives for future research. This article is more about asking questions than answering them, and any answers I offer should be regarded only as suggestion for further discussion.

There is an emerging interest in comparative views of human and nonhuman gesture. Enthusiastic reactions to the workshop on "Gestural communication in human and nonhuman primates" held in Leipzig in the Spring of 2004 revealed this mutual interest. The discussions during the workshop as well as the contributions to this volume indicated that there is still a lot of interesting ground to cover and that a comparative view will enhance understanding of both — gestures in nonhuman AND in human primates. One of the main interests for a comparative view of gestural communication in nonhuman and human primates is that of the evolutionary origins of language. However, comparing gestures in apes and man may also enhance our understanding of the properties of human gestures — for instance, by clarifying what are the evolutionary 'old' forms of gesturing, those which 'survived' the evolution of vocal language, when they appear ontogenetically, and whether their role changes when vocal language is fully developed. On the other hand, knowing which of the gestures 'survived' the evolution of language may shed light on gesture creation and use in nonhuman primates and provide new insights into possible protoforms of language. This article serves as a brief and systematic discussion of more general reasons for a comparison of gestural communication in nonhuman and human primates but also specifying which aspects of gesture can be compared.

The discussion will take as a guideline the questions that were raised in the Introduction — and which were also the 'framing' questions of the workshop:

1. What are the major goals of studies concerning gestural communication in nonhuman and human primates, respectively?
2. To what extent are the gestural communication systems of nonhuman and human primates comparable?
3. Do studies of gestural communication in nonhuman and human primates help to clarify the scenario for the evolution of language?
4. What are the methodological steps with regard to data collection, analysis, and coding?
5. How can a gesture be defined and where does intentionality come into play?
6. What are the semiotic structures of gestures and how do they relate to cognitive processes?
7. What are the structural properties of gestures?
8. What are critical contexts of use and what are the functions of gestures?
9. What kinds of gestures must be distinguished?

What are the major goals of studies concerning gestural communication in nonhuman and human primates, respectively?

Nonhuman primates

It seems that, at present, there are two major perspectives in the analysis of gestural communication of nonhuman primates: a documentary view and an evolutionary view.[1] While the documentary view is primarily concerned with close accounts of forms and functions of behavior and social structures, and adopts a more traditional ethological perspective (Goodall, 1986; Tanner, 2004; van Hooff, 1973), the evolutionary view studies forms of behavior with the aim of identifying possible precursors of language, either in terms of proto linguistic socio-cognitive abilities

(Maestripieri, 1996, 1997 a, 1999; Pika, Liebal, & Tomasello, 2003, 2005; Tomasello & Call 1997; Tomasello et al., 1985, 1989, 1994, 1997) or in terms of basic linguistic capacities (Chalcraft & Gardner, 2005; Gardner et al., 1989; Greenfield & Savage-Rumbaugh, 1990, 1991; Terrace, 1979a, b).

Clearly the evolutionary view (analyzing gestural communication in search for evolutionary trends within the different ape species, as possible precursors of a cultural evolution of language) is currently the prominent influential motivating force. It connects documentation of behavior with one of the big questions of humanity: How did language evolve and which components of the process are ruled by either cultural or genetic evolution.

Human primates

A similar bifold trend characterizes the perspectives taken on gestural communication in human primates: there is a smaller tradition focusing on the documentation of forms and uses of gestures, researchers working from a developmental (Bates et al., 1979; Volterra & Erting, 1990), anthropological (Goodwin, 2000; Streeck, 1993, 1994, 2002), sociological (Schegloff, 1984), sociopsychological (Bavelas et al. 2001; Gerwing & Bavelas, 2004) ethological, semiotic, and linguistic perspective (Calbris 1990; Kendon, 1972, 1980a, 1985, 1988a, 1995, 2002, 2004; Müller, 1998a, b, 2001, 2003, 2004a, b, to appear; Müller & Speckmann, 2002; Müller & Posner, 2004, Neumann 2004; Seyfeddinipur, 2004).

The big trend and the one that has been a strong catalyst for the study of human gesture in the last two decades is related to another major question for humanity: the architecture of the human mind and brain. The cognitive sciences — in their widest sense — have discovered gesture as a 'window onto thought' (Beattie, 2003; McNeill, 1992, 2000, 2005). Gestures during speech are regarded as expressions of thought processes and have thus been studied with regard to the nature of thinking while speaking (imagistic and propositional forms of thought, dynamic thought: McNeill, 2005; metaphoric thinking: Cienki, 1997, 1998, 2000, 2003; Müller, 2004b; Cienki & Müller, to appear), with regard to thinking for speaking and issues of linguistic relativity (adjustment of thinking to language specific structures Haviland, 1993, 1996, 2000; McNeill & Duncan, 2000; Müller, 1998a; Özyürek et al., 2005; Slobin, 1987, 1991, 1996), as well as with respect to processes of learning (Goldin-Meadow, 2003) and to the neuropsychological approach of functional localization of gesture and speech in the brain (Kita & Lausberg, subm.; Lausberg & Cruz, 2004; Lausberg, Ptito, & Zaidel, subm.; Lausberg et al., 2003a, b; McNeill, 1992, 2005). So far, however, there is only one facet within this broad cognitive trend in gesture studies which directly relates the study of human gesture to nonhuman primates — and this is the developmental perspective. Research conducted in this area tends to focus on gestures in preverbal infants — assuming that this is a stage in human development which is most likely to be comparable to nonhuman primate forms and capacities of gestural communication and one that therefore may shed light on a possible gestural origin of human language (for declarative and imperative pointing see Tomasello & Camaioni (1997), pointing in human preverbal children see Pizzuto & Capobianco (2005), Capirci et al. (2005), Liszkowski (2005)).

To what extent are the gestural communication systems of nonhuman and human primates comparable?

This question cannot really be answered until we have gathered much more descriptive knowledge of gestures than is available today. It is interesting to note, however, that in systematic terms more is known about repertoires of gestural communication of nonhuman primates than about the human ones. Primatologists have provided documentations of existing repertoires in different species with accounts of their forms, functions, usages, and of their structural complexity (Liebal, 2005; Liebal, Call, & Tomasello, 2004a; Maestripieri, 1997a, b, 1999, 2005; Tanner, 2004; Tomasello et al., 1985, 1989, 1994, 1997; van Hooff, 1973). For humans such an encompassing documentation on the forms, functions, usages, and structures of gestural communication does not exist; a systematic descriptive study of human gesture as companion or part of speech has not been a foremost aim in current studies; Adam Kendon's recent book "*Gesture. Visible action as utterance*" is a contemporary attempt to cover fundamental characteristics of the phenomenon of human gesture and comes closest to such an encompassing view (Kendon, 2004). Kendon regards gesture as a form of visible action which is used as an utterance or as part of an utterance while McNeill in his influential book "Hand and Mind" takes gesture as a form of visible thought and devotes several chapters to the semantics and discourse function of co-verbal gestures (McNeill, 1992). The only attempt to provide an overview of the entire range of forms and functions, and usage contexts in human gesture is offered by the ethologist Desmond Morris and his colleagues (Morris, 1977, 1994; Morris et al., 1980). Yet this ethological approach (cf. also Cranach & Vine, 1973; Tannen & Saville Troike, 1985; Eibl-Eibesfeldt, 1970; Hinde, 1972) to the analysis of human gestures has not been very influential in contemporary gesture studies (Kendon's work counts as an important exception here), although a close descriptive knowledge of gesture as a medium of communication is clearly highly pertinent for current research.

Researchers of human gestures will benefit from the procedures and the systematicity of descriptions of gestures in nonhuman primates. Based on such an account we may then be able to identify precisely what are possible precursors of human forms of gesturing, where we find overlaps (in terms of evolutionary 'old' forms and uses), where we find differences, and to what extent gestural communication in human and nonhuman primates is in fact comparable.

Do studies of gestural communication in nonhuman primates help to clarify the scenario of the evolution of language?

Productive research into gestural communication within and across different species has vividly shown this is indeed a worthwhile path to follow. Primatologists have argued that gestures are more likely to have functioned as precursors to language because they show a much higher degree of flexibility than vocalizations (Pika et al., 2005). And indeed the anatomic properties of the hands at least in great apes are such that they are able to form signs of sign-language, whereas their vocal apparatus is comparatively unflexible and does not allow for the development of a broad range of different sounds.

Departing from what is known about human forms of gesturing may indicate what to look for in nonhuman primates gesturing and may offer further insights into possible precursors of language; the more systematic knowledge about the nature of human gestures we have (in terms of structures, semiosis, functions, usages) — the better we can compare the communicative and cognitive skills which nonhuman primates display. Empirical investigations of this kind may hence provide substantive arguments in favor or against theories of a gestural evolution of language (Condillac, 1971/1756; Hewes, 1973; Kendon, 1975; Rizzolatti & Arbib, 1998; Roy & Arbib, 2005; and the review of Corballis (Copple, 2005).

What are the methodological steps with regard to data collection, analysis, and coding?

A shared methodology is obviously a high priority for comparative studies, especially since there is no such thing as a developed methodology for the analysis let alone the coding of gestures. As a consequence research results may easily be questioned and may be difficult to replicate. This is especially important because micro-analytic studies of gestures are only possible since video documentation is available (they presuppose the possibility to view a certain gesture repeatedly).

These close observational studies are rooted in a tradition of two millennia of observationally based gesture descriptions with quite a few elaborate accounts of gesture. [2] Despite this tradition, basically all newcomers to the field tend to invent their own steps of analysis, their own transcription and coding conventions, and face similar difficulties often without training from a gesture specialist. Nobody would expect to be able to describe vocal or sign language, talk in interaction, or even facial expression without being a trained linguist, conversation analyst or a psychologist but human gestures seem to be perceived as so transparent that no training, no education seems necessary. I shall argue, instead, that the same kind of expertise is necessary when embarking on the analysis of gestures.

Data collection. In addition, to controlled analysis, we need a systematic consideration of the problems of data collection because data collection significantly but often tacitly steers the analysis. It should be noted: what kinds of situations are recorded (naturalistic or experimental situations, social contexts, number of participants); how many cameras are used (what do they *not* show, how do they prestructure the situation); how do we break down the material (what are the sequences chosen for detailed analysis and why); how are gestural forms transcribed or annotated (not at all, i.e. direct coding, or transcription of form without function, or transcription of function but not form)?

Analysis and coding. At first of course a criterion is needed to distinguish gestures from non-gestures. Then, once a gesture is identified as being gesture, the analytic steps followed must be spelled out: hence, how exactly do we arrive at a functional, semantic, or pragmatic analysis of a gesture. And, when coding is afforded, then what are the criteria for inclusion into one or the other coding category. Note, that any coding decision presupposes two steps: (1) identification of what counts as a gesture (or as two or three gestures) and (2) analysis of form and function. Based on these two analytic steps a gesture is coded. In short, coding presupposes analysis and this analysis must be made explicit — results obtained otherwise become highly questionable.

Kendon (2004) has outlined the following steps of gesture analysis: begin with the identification of what is a gesture (in terms of gesture units and gesture phrases) and then apply a 'context-of-uses' analysis (for a clear exposition of one aspect of Kendon's methodology see Wilkins, 2006, for the criteria of gesture identification see Kendon, 2004, Chapters 2 and 7). Müller (2004a) has sketched an analytic procedure for recurring gestural forms which combines semiotic analyses with context-sensitive sequential analyses.

The next section may be taken as a discussion of some fundamental questions which arise when facing the first step in the analysis: the identification of what is considered a gesture.

How can a gesture be defined and where does intentionality come into play? A suggestion for further discussion

I aim at a definition that is based on formal characteristics of the movement rather than on inner states such as intentionality or communicative intent. Why? Because 'inner' states are only detectable through 'outer' cues. This is a trivial but crucial property because without them, inner states would remain uniquely private states. In order to be communicable they must be made publicly available, in other words they must be perceptible, be it visibly, audibly or tactily. This public availability is in the first place a participant directed property, a property of the movement which makes it recognizable AS communicative for other participants in the interactive encounter but it may of course be exploited systematically by the analyst. I suggest accordingly that a gesture may be defined in terms of three formal properties: the *voluntary execution* of the movement, its *address* and its *sequential position* within the flow of surrounding activities.

Voluntary execution of the movement relates to formal features of movements which make them perceptible as controlled rather than un-controlled movements of the limbs.[3]

Gestures would be movements that are "under voluntary control" performed in the absence of a direct practical aim.

This helps to distinguish practical actions such as holding a glass of water and moving it to the mouth with the practical aim of drinking, from as-if-actions, such as holding an empty glass and moving it to the mouth with the possible communicative aim of indicating a desire to drink; or from as-if-actions where the hand is held as-if-holding a glass and moved to the mouth as-if-moving a glass to the mouth, with the communicative aim of indicating that the person wants something to drink, or of describing another person who is drinking, or whatever may be the locally established communicative intent.[4]

Voluntary control of the movement provides the basis for the distinction between gestural and symptomatic behavior. As Kendon (2004) points out the term gesture is usually used to refer to such actions as waving good bye, greetings, pointing or pantomimic enactments of actions, or descriptions of shapes of objects. It is usually not used to refer to such behavior as laughter, crying, blushing, trembling, nor is it used to refer to self-adjustments of clothes or self-grooming. Note that a similar intuition underlies the exclusion of non-flexibly used body movements and exdpression of inetrnal, emotional states in Tomasello's research group.

Address of a movement captures the communicative directedness towards a co-participant.[5] Address can be made visible by directing the movement itself or by gazing before, during, and/or after the performed movement at an addressee. Orientation of the movement might be a direct result of the bodily orientation towards a co-participant (somebody directly facing another person does not have to make any effort to gesture in the direction of this person) or it might need extra effort to be noticed (when somebody is positioned at a 90 degree angle of a co-participant or even outside the visual field of a potential addressee).

Sequential position within the flow of surrounding activities is a third formal feature which is crucial in determining of what is a gesture.

One of the core criteria that has been used for the identification of intentional gestures is 'response waiting'. Here is an explication from Tomasello and colleagues: "When the juvenile directed a conventional act at another animal and then clearly waited for a response, for example, slapping the ground at a peer and waiting and watching in a crouched position, we were much more confident that the act was indeed illocutionary." (Tomasello et al., 1985, pp. 179).[6] 'Response waiting' is determined through its sequential position: 'waiting for a response' — is an activity which is defined through its position after an initiating activity, an activity which may be taken as a first move in an interactive encounter. Thus waiting for a response in spoken conversation occurs only after a specific other activity such as a question or a request. It is a first part of a bi-fold interactive activity, whereby the second pair part has to be produced by another person: question — answer, offer — rejection, or agreement, greeting — greeting, and the like. Conversation analysts have termed these interactive pairs adjacency pairs and have studied their obligatory structure (Sacks, Schegloff, & Jefferson, 1974). In primate gesturing it has been observed that slaps on the ground may be used as an invitation to play or as an aggressive move in an agonistic encounter. Here too not the features of the gesture as such determine its function but its position in relation to prior activities.

Producing a movement in such a way that it meets all or some of these features may be regarded as a form of communicative competence, which may well function as an evolutionary precursor of language. Take for instance the capacity of different species to detach their visual attention from the food source to establish visual contact with a person who is in a position to reach the food. I am referring here to an experiment in which evolutionary younger chimpanzees and bonobos were compared with evolutionary older gorillas and orangutans (Liebal, Pika, Call and Tomasello, 2004b). It was found that in experimental situations when interacting with humans in a food begging task, both chimpanzees and bonobos will actively seek the visual attention of the human even if the food is located behind the experimentor and therefore separated from him whereas gorillas and orangutans do not seek visual attention.[7]

Note, that these formal features are mutually reinforcing, in the sense that when they are all present — there is strong evidence that we may indeed characterize a bodily movement as gesture; when they do not overlap, as in instances where slapping a stick on the ground as a possible request for play, while the bodily orientation of the chimpanzee is directed away from its presumed partner, than the situation is less clear; when in this context however, the action is repeated and probably gaze is directed to the possible partner then this may serve as a further indication to characterize it as a gesture. We will have to accept, however, that the boundaries between a gesture, a practical action, a self-directed 'monological' activity, and an involuntary expression are fuzzy. Yet with the proposed set of criteria at hand we are in a position to determine 'the degree of fuzziness' more precisely.

What are the semiotic characteristics of gestures (and how do they relate to cognitive processes)?

It is worthwhile considering more closely what I term the 'semiotic structures of gestures'. Take for example the characterization of a gesture as iconic. This seems pretty straightforward and clear — yet iconicity may be realized in many different ways (Kendon, 1980b, c; Kendon 1988b; Mandel, 1977; Müller, 1998a, b; Taub, 2001; Wundt, 1921). In humans iconic gestures may reenact an action such as offering something on the open hand, opening a window, or throwing a bowling ball; they may also model objects such as the rectangular shape of a window or the round shape of the bowling ball; or they may outline objects with the fingers as if drawing with a pencil; furthermore these objects may be represented through the shape of hands themselves: when the window is represented as a flat object through the flat extended hand, or the bowling ball through a fist shaped hand (examples are taken from my data). Müller has proposed to term these different forms of iconicity in gesture 'gestural modes of representation' (Müller 1998a, b) and distinguishes four basic principles of creating an iconic gesture: the hand imitates, the hand models, the hand draws, the hand embodies. This fourfold distinction is not conceived as an extensional characterization of all possible forms of iconic semiosis — rather it is intended as a first distinction of basic principles. There are indications that these principles may combine or may need to be further differentiated — but these are issues for future research.

In addition, semiotic structure in gesture can not be reduced to iconicity. A second core semiotic structure is deixis.[8] Deixis being conceived roughly as establishing reference to something by indicating where it is located rather than representing properties of the action or object in play (holding for space, time, and person): the most prominent deictic gesture is pointing with the extended index finger; but as we know there are different possible forms of pointing (the flat extended hand, the head, the thumb, the eyes) (these are intracultural variations, cf. Kendon, 2004; Kendon & Versante, 2003) or by protruding the lips (these are cross-cultural variations, cf. Enfield, 2001; Sherzer, 1973); we also know that deictic and iconic gestures may fuse; as for instance when objects are localized in the gesture space and serve as landmark for further verbal and gestural explanation (Fricke, 2002, 2004), or when contrasting arguments are placed in the gesture space in opposite loci (Calbris, 2004), or when pointing is combined with drawing the shape of the object referred to (Kendon, 2004). When deictic gestures are not reduced to pointing with the extended index finger, then this offers a new perspective on potentially proto-deictic forms in gestural communication of nonhuman primates.

Based on a better understanding of what are iconic and what deictic features in gestures we may then be in a position to determine more precisely what Roy and Arbib (2005) mean when they say: "In the context of the mirror system hypothesis, it seems that the most evolved communicative gestures in non-human primates take the shape of deictic movement" (see also Pika et al., 2005). Such a distinction will also help us understand better Arbib and Roy's conclusion that iconicity in the gestures of nonhuman primates is basically absent.[9] To me it rather seems that at least certain kinds of iconicity are present in gestures of nonhuman primates, namely the one that belongs in the first mode of representation: the hand imitates actions through modulation (cf. Goffman, 1974; Müller & Haferland, 1997). Modulation follows an iconic principle in that the modulated action resembles the full-fledged action in important respects — namely the meaningful traits of the action. If an action like receiving food on the flat extended hand is to be turned into a gesture (indicating the wish to receive food or the wish to play chasing), then it is necessary to know what are the 'meaningful' parts of the action and which ones, can be dropped or left out entirely, which ones can be abbreviated so that the movement is still recognizable as modulated action. If a flat extended hand is to be understood as a request for receiving food, the place where the hand is located is indicative of the food source: if it is the mouth of the mother, then placing the hand below the chin of the mother makes sense, if it is a fruit on the ground then moving the hand there would make sense. If on the other hand the orientation of the palm is changed (not palm up but palm down) this also changes the meaning of the gesture significantly; the gesture would presumably not be functional any more as a request or offer gesture — but this is at present still an open empirical question. I wish to suggest, that every mode of representation implies a specific form of cognitive activity, a different kind of abstraction from the perceived event or object. Meaningful traits of the action or of an object must be identified and abstracted from a perceptional gestalt. It might turn out that iconicity, which is based on the abstraction of a perceived object rather than on the abstraction from actions, is absent in apes — but this at present is but a guess, another starting point for further comparative studies.

To conclude: a more differentiated view of iconicity would enable us to identify more precisely whether nonhuman primates use iconic gestures or not and if they do then what are the cognitive processes involved in these gestures. It would moreover allow us to determine more precisely the forms of iconicity that evolved in the context of human language. And this would in turn contribute to a better understanding of the vividly disputed role of iconicity in language (Fischer & Nänny 2001; Maeder, Fischer, & Herlovsky, 2005; Müller & Fischer, 2003; Nänny & Fischer, 1999; Posner, 1994, 2003).

What are the structural properties of gestures?

Structural properties of gestures have barely been investigated at all. This is striking because it is rather obvious that gestures vary significantly with regard to their simultaneous and linear structure. This distinction is based on the parameter of time, simultaneous structure refers to temporarily parallel and linear structure refers to a temporal sequence.

Simultaneous structure. Gestures may differ with regard to their constitutive features: configuration, position, movement, and orientation. In other words a gesture is at one and the same moment in time structured in terms of these four features. In some gestures these features may systematically vary, i.e., when for instance the same configuration (palm up and open) is used in different positions or locations, with the location changing the meaning of the gesture significantly. Or the same configuration may be reduplicated by using both hands instead of one hand (parallel structure), or a one hand movement may be repeated (linear structure) — both are iconic forms of intensification achieved through structural means. Kendon has based his concept of a gesture family on the recurrence of such features (cf. Kendon 2004; Müller 2004a). To which degree these combinations are flexibly combined and to which degree they are holistically composed remains to be clarified through systematic empirical analyses of these forms. There are some indications that a combinatoric structure is to be found in highly recurrent gestural forms — as for instance the palm-open-hand or the ring-hand — whereas spontaneously created gestures which depict a specific object or person in a specific situation — a window falling down from a building, a friend playing violin, or the shape of a picture frame in the living room — seem to be more structured in terms of gestalts.

Another form of parallel structure in gesture is the combination of different body parts. As far as I can see research in human gestures focuses almost entirely on hand gestures (exceptions are Kendon, 2002; McClave, 2000); yet we know that not only may hand gestures be substituted by head gestures and vice versa (Negation with head shake and negation with lateral hand or finger movement; cf. Kendon, 2002; Becker, 2004) but we also know that hand gestures may combine with head or shoulder movements, or may involve the arms and the upper body part. Again a systematic account of this structural aspect of gestures awaits detailed investigation — for a comparative view this appears a rather important issue, since in nonhuman primates gestures involving other body parts than the hands seem to be frequent. How important and frequent the involvement of other body parts is in human hand gesturing, is at the moment still unclear.

Linear structure. Gestures also combine on the linear level, i.e. in temporal sequences. In nonhuman primates clapping hands or slapping the ground is used as attention getter for a subsequent invitation to play (Liebal et al., 2004a). Humans may combine gestures semantically when for instance a first gesture depicts a picture frame and a second gesture a crown which sits on top of it; or when somebody outlines a huge pot gesturally and then performs a gesture which takes something out of this virtual pot. In contrast, sequential combination in chimpanzee gestures do not lead to a modification of the meaning of the gestures; sequences mostly occur in contexts of no response, i.e. are repetitions or intensifications (Liebal, 2005; Liebal et al., 2004a). Furthermore, in human gesturing, we find complex scenarios of events (a visit in a park with beautiful weather, wind and rain) and we find, what could tentatively be termed 'local conventionalization' i.e. one and the same form of a gesture attains a specific meaning within and for one conversation. These are cases which resemble the phenomenon of ontogenetic ritualization (Tomasello et al., 1985, 1989, 1994, 1997), where a certain gesture attains a fixed meaning in a mother-child dyad, but remains in this dyad.

A comparative view of structural properties of gestures would be revelatory for both sides, not only because it touches on the question of what is regarded a gesture but because the degree to which simultaneous and linear structuring of gesture is present in nonhuman as compared to human primates may offer further insights into proto-forms of linguistic structure. Hence for instance, in terms of parallel structure it is possible that liberating the hand from other body parts may have constituted a similar important evolutionary step as the capacity to combine gestures in linear structures, which is a strong prerequisite for syntactic structuring.

What kinds of gestures must be distinguished?

This question touches upon the nontrivial question of classification of gestural movements.[10] Quite a few classification systems have been proposed for human gestures — but not for nonhuman gestures. The question of how to classify gestures is a theoretical challenge and so far no satisfying proposal has been put forward. The systems offered differ in various respects from each other: some of them focus on hand gestures used along with and in relation to speech uniquely (Efron 1941[1972]; Freedman, 1977; McNeill, 1992; Müller, 1998a) whereas others include facial expression of affect, self-adaptors, and regulators of social interaction (Ekman & Friesen, 1969), some of them include self-touching movements (Freedman, 1977; Müller, 1998a) others exclude them. It is important to bear in mind that all proposals classify gestures with relation to a specific goal and against a specific background (cf. Kendon, 2004). Efron needed to find categories which would allow him to compare the gestures of two different groups of immigrants and two groups of their descendants in such a way that it would be possible to see whether their gestures are subject to culturally motivated changes; Freedman related his categories to the psychic processes of focusing and representing and based a psychoanalytic theory of gestural communication on these two psychic structures present in what he termed body-focused (i.e., hands touching the body) and object-focused movements (i.e., hands moving in front of the body). McNeill in turn relates his categories to different aspects of speaking and thinking: hence iconic gestures represent a concrete entity, while metaphoric gestures represent an abstract one, deictics point towards a concrete or abstract referent, beats and cohesives are different ways of structuring the discourse. Because McNeill is interested in gestures as direct expression of thought processes he introduces four further criteria to narrow down his focus in classifying gestures: relationship to speech, to linguistic properties, to conventions, and character of semiosis (McNeill, 2000, pp. 2–7). With reference to what he termed 'Kendon's continuum' (McNeill, 1992, p. 37) he focuses in his work (and accordingly in his classification) on gestures which are companions of speech, which do not have linguistic properties, which are not-conventionalized and which show a global and synthetic semiotic structure. He narrows down his focus on these kinds of gestures because they are the ones that allow a direct view onto actual online thought processes.

For a comparative view it would be very useful to have well-grounded classification systems of gestures (based on the same classification principle for every category) because it would then be easy to see which forms are present and which ones are not. Departing from a functional classification principle (cf. Müller, 1998, 1999) — would probably reveal that performative gestures which are communicative actions often directly derived from practical actions (such as begging, presenting, offering with the Palm-Up-Open-Hand, cf. Müller & Haferland, 1998) are present in both humans and nonhumans, whereas referential gestures (gestures which are used to refer to some object, event or entity) are widely present in human gestures and only scarcely present (if at all) in nonhuman gestures (see Pika et al., 2005; Gomez, 2005). Classifying gestures always relates to a specific research interest; changing the perspective and the question will change the range of phenomena to be considered and will require to work out a new classification (cf. also Kendon, 2004). If for example, conventionality is used as a criterion we may find that a continuum of more or less conventionalized forms of gestures exists in humans, whereas apes only very rarely conventionalize gestures spontaneously but are ready to use conventionalized gestures (such as pointing or even Sign Language) when exposed to them in a human environment. When contrasting conventionalization with ontogenetic ritualization (as Tomasello and colleagues do) it may turn out, that both are present in humans, yet whether spontaneous conventionalization as a social process which transcends the mother–child dyad is indeed present in nonhuman primate communication continues to be a hot issue in primatology.

What are critical contexts of use and what are the functions of gestures?

The contexts in which gestures are used constitute a core point of reference in studies of nonhuman primates' forms of gesticulation (cf Tomasello et al., 1985; Liebal, 2005) in contrast, they do not play a significant role in the study of human gestures. In primate research usage contexts are documented in terms of social contexts such as aggression, mating, affiliation, or parental care (e.g., Tomasello et al., 1985, 1987), or more specific aspects of social interaction such as cooperation and competition (Maestripieri, 1997b). Functions of gestures tend to be determined in relation to these various contexts.

It is striking that in studies of human forms of gesticulation contexts in which gestures are used have not been systematically documented. It seems however, that certain discourse contexts are related to certain kinds of gestures; McNeill's account of gestures with concrete and abstract reference in gestures shows that narratives of a cartoon will raise a lot of gestures depicting concrete entities and activities, whereas discourse on mathematical topics will trigger gestures which depict abstract concepts, while both contexts may now and then also stimulate gestures which relate to the narrative or discourse structure (cf. McNeill, 1992;

see also Müller, 2003); a case study indicates that for instance a heated argument triggers numerous gestures that present, reject, challenge or evaluate what is said while gestures used to describe concrete or abstract objects or activities are relatively rare — but so far these are only more or less accidental observations and a systematic inquiry of the relation between contexts of usage and functions of gestures would provide interesting insights into the nature of gesture as a medium of communication. It would show whether there are specific contexts of use which are specifically 'gesture productive' while others are less and how they relate to communicative functions of gestures.

This does not imply however, that gestures only have one function at a time. It has been argued that human gestures may have different functions in parallel (Bühler, 1934; Müller, 1998a). A gesture may display the affective state of the individual (in terms of the quality of the movement), it may represent some object or action in the outer world (a pot, opening a window, asking for an object with the extended open hand), and it is addressed to a recipient — a co-participant in the interaction. This concept of functional dimensions goes back to Bühler's (cf. Bühler, 1934; Müller, 1998a) classical distinction of three dimensions in the verbal sign. Bühler distinguishes 'expression', 'representation' and 'appeal' ('Ausdruck', 'Darstellung', 'Appell'): 'expression' is the dimension of a verbal sign that relates to inner states (in this regard the verbal sign functions as symptom); 'representation' relates the sign to entities and events in the world (in this respect it functions as a symbol); 'appeal' characterizes the fact that any verbal sign is inherently communicative, i.e. directed towards an addressee (in this sense the verbal sign functions as signal).

In terms of a comparing contexts of use and functions of gestural communication in human and nonhuman primates it would be interesting indeed to see whether and if yes to what degree such a multifunctionality is observable. Since not all gestures (also in humans) have a representational function, it is clear that all three functions in parallel will not be present unanimously. But what about the expressive and the appealing function — are they present in parallel? If so, in what form? If they are present then could it be that they are evolutionary older forms of communication, forms of communication that provide the scaffolding for the emergence of intersubjectivity as in such early forms of joint attention as was documented for early child communication (Bates, 1979; Capirci et al., 2005; Pizzuto & Capobianco, 2005). An interesting hypothesis could then be that in evolutionary terms these functions have evolved sequentially and only merged at a later stage — which would mean that language capacity would presuppose the capacity of layering communicative functions

Comparing gestural communication in nonhuman and human primates: Concluding remarks

Hitherto gestural communication has not been studied intensely from a comparative point of view. Somehow it seems as if gesture scholars tacitly assume that there is not much to compare really. The only exceptions are the above mentioned developmental studies which have influenced primatologists' analysis of pointing in apes to a certain degree (declarative and imperative pointing, Volterra & Erting, 1990). But apart from this one — at least as far as I can see — there is no line of research of nonhuman and human gestures really, which takes a comparative view as a point of departure. I hope to have shown, however, that this constitutes a very fruitful perspective for future research. More specifically I hope to have indicated that quite a few questions become perceptible only when departing from a descriptive approach. As I have mentioned already the analysis of gestures in humans tends to be easily regarded as unproblematic, gestures being conceived of as transparent, crystal clear windows onto thought. Accordingly, human gestures are often treated as something we 'all understand' without further expertise; this is, I believe, a highly problematic assumption, given the factual complexity of forms of gestural communication. Primatologists who cannot rely on verbal context in which a gesture is produced have not been tempted to conceive of gestures as transparent windows onto cognitive processes. Moreover, I suggest that a descriptive point of departure is crucial also for experimental studies because the deductions of cognitive, social, affective, communicative skills are only as good as our knowledge of the gestural signs on which we base our inferences.

This also holds with regard to the big questions of the evolution of language and the architecture of the mind and the brain. A close documentation of the nature of gesture (and speech) may help to underline that structure in language is much more than syntactic structure — we find structures at many levels. If we can find out which of them are present in nonhuman primates this may shed more light onto a possible gestural origin of language. It may put us into a position to reconstruct more exactly the evolutionary steps that might have changed Condillac's *langage d'action* into a combination of gestures and unarticulated vocalizations (*gestes et cris inarticulés*) to a full fledged language with gesture as an integrated part.

Notes

* I am most grateful to Adam Kendon and Katja Liebal for invaluable critical and inspiring comments on earlier versions of this paper.

1. I am using the term 'documentary view' in the sense that anthropological linguists use it for the documentation of non-written languages. I believe that adopting a documentary view with regard to gesture is a highly important task for future research and a crucial precondition for elaborated cognitive studies of gestures.

2. For an overview see Kendon (2004) and Müller (1998).

3. Kendon (1978, 1985, 2004) reports on a study which suggests that humans are readily able to decide whether a movement they see is a gesture or not — even when they do not have access to sound; yet the exact formal properties of 'voluntary execution' remain to be identified. See Kendon (2004) for a detailed discussion of this issue and Kendon and Müller (2000) for a sketch of what is to be regarded a gesture.

4. See Müller (1998, 1999) for a detailed account of gestures as 'as-if-actions'.

5. I am following Kendon's use of the term 'address' here [personal communication]. For the role of bodily movement in social interaction see Kendon (1970, 1972, 1973) and Chapter 7 of Kendon (2004).

6. For further criteria of intentional behavior see Bates et al. (1979) and Liebal et al. (2004).

7. Primatologists have investigated these kinds of phenomena in the context of adjustments to audience effects. See also Pika et al. (2005).

8. It is a worthwhile aside, that Karl Bühler (1934) regards pointing ('Zeigen') and naming ('Benennen') as THE two realms of language — more specifically of the representational function of language and he is the first to systematically include deixis into a theory of language.

9. See Tanner and Byrne (1996) for the opposite view.

10. See for a critical overview and discussion of classification systems from Antiquity to the Present Kendon (2004).

References

Bates, Elizabeth, Laura Benigni, Inge Bretherton, Luigia Camaioni, & Virginia Volterra (1979). *The emergence of symbols: Cognition and communication in infancy.* New York: Academic Press.

Bavelas, Janet & Christine Kenwood, Trudy Johnson, & Bruce Phillips (2001). An experimental study of when and how speakers use gestures to communicate. *Gesture 2* (1), 1–17.

Beattie, Geoffrey (2003). *Visible thought. The new psychology of body language.* London & New York: Routledge.

Becker, Karin (2004). Zur Morphologie redebegleitender Gesten. Unpublished MA Thesis. Berlin: Free University.

Blake, Joanna, Grace Vitale, Patricia Osborne & Esther Olshansky (2005). A cross-cultural comparison of communicative gestures in human infants during the transition to language. *Gesture, 5*, 201–217. (This volume)

Bühler, Karl (1934/1990). *Sprachtheorie. Die Darstellungsfunktion der Sprache.* Jena: Fischer [*Theory of Language: The Representational Function of Language*, trans. Donald Fraser Goodwin (Foundations of Semiotics, 25). Amsterdam & Philadelphia: John Benjamins, 1990].

Calbris, Geneviève (1990). *The semiotics of French gestures.* Bloomington: Indiana University Press.

Calbris, Geneviève (2004). Déixis représentative. In Cornelia Müller & Roland Posner (Eds.), *The semantics and pragmatics of everyday gestures. Proceedings of the Berlin conference April 1998* (pp. 145–156). Berlin: Weidler Buchverlag.

Capirci, Olga, Annarita Contaldo, M. Cristina Caselli & Virginia Volterra (2005). From action to language through gesture: A longitudinal perspective. *Gesture*, *5*, 155–177. (This volume)

Chalcraft, Valerie J. & R. Allen Gardner (2005). Cross-fostered chimpanzees modulate signs of American Sign Language. *Gesture*, *5*, 107–132. (This volume)

Cienki, Alan (1997). Motion in the metaphorical spaces of morality and reasoning as expressed in language and gesture. *International Journal of Communication. A Review of Cognition, Culture & Communication,* 7(1/2), 85–98.

Cienki, Alan (1998). Metaphoric gestures and some of their relations to verbal metaphorical expressions. In Jan-Pierre Koenig (Ed.), *Discourse and cognition: Bridging the gap* (pp. 189–204). Stanford, CA: Center for the Study of Language and Information.

Cienki, Alan (2000). The production of metaphoric gestures: Evidence for on-line cognitive processing of metaphoric mappings. Ms.

Cienki, Alan (2003). Overview of, and potential for, research on metaphor and gesture. Introduction to the theme session of metaphor and gesture. ICLC 2003. Ms.

Cienki, Alan & Cornelia Müller (to appear). Metaphor and gesture. In Raymond W. Gibbs (Ed.), *Cambridge handbook of metaphor and thought.* Cambridge: Cambridge University Press.

Goldin-Meadow, Susan (2005). The two faces of gesture: Language and thought. *Gesture*, *5*, 241–257. (This volume)

Gómez, Juan Carlos (2005). Requesting gestures in captive monkeys and apes: Conditioned responses or referential behaviours?. *Gesture*, *5*, 91–105. (This volume)

Condillac, Etienne Bonnot de (1971/1756) *An essay on the origin of human knowledge (1756).* Facsimile reproduction of the translation of Thomas Nugent, edited and with an introduction by Robert G. Weyant. Delmar, New York: Scholars' Facsimiles and Reprints.

Copple, Mary (2005). Review of Michael C. Corballis, 2002, From hand to mouth. The origins of language. Princetown, Oxford: Princetown University Press. *Gesture*, *5*, 285–304. (This volume)

Efron, David (1941[1972]). *Gesture, race and culture.* Paris, The Hague: Mouton.

Ekman, Paul & Wallace V. Friesen (1969). The repertoire of nonverbal behavior: Categories, origins, usage, and coding. *Semiotica,* 1(1), 49–98.

Eibl-Eibesfeldt, Irenäus (1970). *Human ethology. Biology of Behavior.* (Translated from German). New York: Holt Rinehart and Winston.

Enfield, Nick (2001). 'Lip-pointing': A discussion of form and function with reference to data from Laos. *Gesture,* 1(2), 185–212.

Fischer, Olga & Max Nänny (Eds.) (2001). *The motivated sign. Iconicity in language and literature* (Vol. 2). Amsterdam & Philadelphia: John Benjamins.

Fricke, Ellen (2002). Origo, pointing, and speech — The impact of co-speech gestures on linguistic deixis theory. *Gesture*, 2(2), 207–226.

Fricke, Ellen (2004). Origo, Geste und Raum. Lokaldeixis im Deutschen. Unpublished Doctoral Thesis. Berlin: Technical University.

Gardner, Beatrix T., R. Allen Gardner, & Susan G. Nichols (1989). The shapes and uses of signs in a cross-fostering laboratory. In R. Allen Gardner, Beatrix T. Gardner, & Thomas E. van Cantfort (Eds.), *Teaching sign language to chimpanzees* (pp. 55–180). Albany, NY: SUNY Press.

Gerwing, Jennifer & Janet Bavelas (2004). Linguistic influences on gesture's form. *Gesture*, 4(2), 157–195.

Goldin-Meadow, Susan (2003). *Gesture. How our hands help us think*. Cambridge: Harvard University Press.

Goffman, Erving (1974). *Frame analysis*. New York: Harper.

Goodall, Jane (1986). *The chimpanzees of Gombe: Patterns of behavior*. Cambridge: Harvard University Press.

Goodwin, Charles (2000). Gesture, aphasia, and interaction. In David McNeill (Ed.), *Language and gesture*. Cambridge: Cambridge University Press.

Greenfield, Patricia Marks & E. Sue Savage-Rumbaugh (1990). Grammatical combination in *Pan paniscus*: Processes of learning and invention in the evolution and development of language. In Sue Taylor Parker & Kathleen Rita Gibson (Eds.), *"Language" and intelligence in monkeys and apes* (pp. 333–355). Cambridge: Cambridge University Press.

Greenfield, Patricia Marks & E. Sue Savage-Rumbaugh (1991). Imitation, grammatical development, and the invention of protogrammar by an ape. In Norman Krasnegor, Duane Rumbaugh, Michael Studdert-Kennedy, & Richard Schiefelbusch (Eds.), *Biological and behavioral determinants of language development* (pp. 235–258). Hillsdale, NJ: Lawrence Erlbaum Associates.

Haviland, John B. (1993). Anchoring, iconicity, and orientation in Guugu Yimithirr pointing gestures. *Journal of Linguistic Anthropology*, 3, 3–45.

Haviland, John B. (1996). Projections, transpositions, and relativity. In John J. Gumperz & Stephen C. Levinson (Eds.), *Rethinking linguistic relativity* (pp. 271–323). Cambridge: Cambridge University Press.

Haviland, John B. (2000). Pointing, gesture spaces, and mental maps. In David McNeill (Ed.), *Language and gesture*. Cambridge: Cambridge University Press.

Hewes, Gordon W. (1973). Primate communication and the gestural origin of language. *Current Anthropology*, 14, 5–24.

Hinde, Robert A. (1972). *Nonverbal communication*. Cambridge: Cambridge University Press.

Kendon, Adam (1970). Movement coordination in social interaction. *Acta Psychologica*, 32, 1–25.

Kendon, Adam (1972). Some relationships between body motion and speech: An analysis of an example. In A. W. Siegman & B. Pope (Eds.), *Studies in dyadic communication* (pp. 177–210). New York: Pergamon Press.

Kendon, Adam (1973). The role of visible behaviour in the organization of social interaction. In Mario von Cranach & Ian Vine (Eds.), *Social communication and movement: Studies of interaction and expression in man and chimpanzee* (pp. 29–74). London: Academic Press.

Kendon, Adam (1975). Gesticulation, speech, and the gesture theory of language origins. *Sign Language Studies*, 9, 349–373.

Kendon, Adam (1980a). Gesticulation and speech: Two aspects of the process of utterance. In Mary Ritchie Key (Ed.), *Nonverbal communication and language* (pp. 207–227). The Hague: Mouton.

Kendon, Adam (1980b). A description of a deaf-mute sign language from the Enga Province of Papua New Guinea with some comparative discussion. Part I: The formational properties of Enga signs. *Semiotica,* 31(1/2), 1–34.

Kendon, Adam (1980c). A description of a deaf-mute sign language from the Enga Province of Papua New Guinea with some comparative discussion. Part II: The semiotic functioning of Enga signs. *Semiotica,* 31(3/2), 81–117.

Kendon, Adam (1985). Some uses of gesture. In Deborah Tannen & Muriel Saville-Troike (Eds.), *Perspectives on silence* (pp. 215–234). Norwood N.J.: Ablex Publishing Corporation.

Kendon, Adam (1988a). How gestures can become like words. In Fernando Poyatos (Ed.), *Cross-cultural perspectives in nonverbal communication* (pp. 131–141). Toronto: C.J. Hogrefe.

Kendon, Adam (1988b). *Sign languages of Aboriginal Australia: Cultural, semiotic and communicative perspectives.* Cambridge: Cambridge University Press.

Kendon, Adam (1995). Gestures as illocutionary and discourse structure markers in Southern Italian conversation. *Journal of Pragmatics,* 23, 247–279.

Kendon, Adam (2002). Some uses of the head shake. *Gesture,* 2 (2), 147–182.

Kendon, Adam (2004). *Gesture. Visible action as utterance.* Cambridge, Cambridge University Press.

Kendon, Adam & Cornelia Müller (2000) Introducing GESTURE. *Gesture, 1*(1), 1–7.

Kita Sotaro & Hedda Lausberg (in press). Speech-gesture discoordination in split brain patients' left-hand gestures: Evidence for right-hemispheric generation of co-speech gestures. *Cortex.*

Lausberg, Hedda & Robyn F. Cruz (2004). Hemispheric specialisation in the imitation of hand positions and finger configurations. A controlled study in split-brain patients. *Neuropsychologia, 42,* 320–334.

Lausberg, Hedda, Robyn F. Cruz, Sotaro Kita, Eran Zaidel, & Alain Ptito (2003a). Pantomime to visual presentation of objects: Left hand dyspraxia in patients with complete callosotomy. *Brain, 126,* 343–360.

Lausberg, Hedda, Sotaro Kita, Eran Zaidel, & Alain Ptito (2003b). Split-brain patients neglect left personal space during right-handed gestures. *Neuropsychologia, 41,* 1317–1329.

Lausberg, Hedda, Alain Ptito, & Eran Zaidel (in press). Left-hand preference for co-speech gestures in patients with complete callosal disconnection. *Neuropsychologia.*

Leavens, David & William Hopkins (2005). Multimodal concomitants of manual gesture by chimpanzees (Pan troglodytes): Influence of food size and distance. *Gesture, 5,* 75–90. (This volume)

Liebal, Katja (2005). Social communication in great apes. Unpubl. PhD. Thesis. Leipzig.

Liebal, Katja, Josep Call, & Michael Tomasello (2004a). The use of gesture sequences by chimpanzees. *American Journal of Primatology, 64,* 377–396.

Liebal, Katja, Simone Pika, Josep Call, & Michael Tomasello (2004b). To move or not to move: How great apes adjust to the attentional states of others. *Interaction Studies, 5,* 199–219.

Liszkowski, Ulf (2005). Human twelve-month-olds point cooperatively to share interest with and helpfully provide information for a communicative partner. *Gesture, 5,* 135–154. (This volume)

Maeder, Constantino, Olga Fischer, & William J. Herlovsky (Eds.) (2005). *Outside in — inside out. Iconicity in language and literature* (*Vol. 4*). Amsterdam & Philadelphia: John Benjamins.

Maestripieri, Dario (1996). Gestural communication and its cognitive implications in pigtail macaques (*Macaca nemestrina*). *Behaviour, 133* (13/14), 997–1022.

Maestripieri, Dario (1997a). The evolution of communication. *Language & Communication, 17*, 269–277.

Maestripieri, Dario (1997b). Gestural communication in Macaques: Usage and meaning of nonvocal signals. *Evolution of communication, 1*, 193–222.

Maestripieri, Dario (1999). Primate social organization, gestural repertoire size, and communication dynamics: A comparative study of macaques. In Barbara J. King (Ed.), *The evolution of language: Assessing the evidence from nonhuman primates* (pp. 55–77). Santa Fe: School of American Research.

Maestripieri, Dario (2005). Gestural communication in three species of macaques (*Macaca mulatta, M. nemestrina, M. arctoides*): Use of signals in relation to dominance and social context. *Gesture, 5*, 57–73. (This volume)

Mandel, Mark (1977). Iconic devices in American Sign Language. In Lynn A. Friedman (Ed.), *On the other hand. New perspectives on American Sign Language* (pp. 57–107). New York: Academic Press.

McClave, Evelyn (2000). Linguistic functions of head movements in the context of speech. *Journal of Pragmatics, 32*, 855–878.

McNeill, David (1992). *Hand and mind: What gestures reveal about thought*. Chicago: University of Chicago Press.

McNeill, David (Ed.) (2000). Introduction. In David McNeill (Ed.), *Language and gesture*. Cambridge: Cambridge University Press.

McNeill, David (2005). *Gesture and thought*. Chicago: Chicago University Press.

McNeill, David & Susan D. Duncan (2000). Growth-points in thinking-for-speaking. In David McNeill (Ed.), *Language and gesture*. Cambridge: Cambridge University Press.

Morris, Desmond (1994). *Manwatching*. London: Jonathan Cape.

Morris, Desmond (1994). *Bodytalk. The meaning of human gestures*. New York: Crown Publishers.

Morris, Desmond, Peter Collett, Peter Marsh, & Marie O'Shaughnessy (1980). *Gestures. Their origins and distribution*. New York: Stein & Day Publishers.

Müller, Cornelia (1998a). *Redebegleitende Gesten. Kulturgeschichte — Theorie — Sprachvergleich*. Berlin: Verlag Arno Spitz.

Müller, Cornelia (1998b). Iconicity and gesture. In Serge Santi et al. (Eds.), *Oralité et Gestualité: Communication multimodale, interaction* (pp. 321–328). Montréal, Paris: L'Harmattan.

Müller, Cornelia (1999). Lectures on Gesture. University of Chicago. Ms.

Müller, Cornelia (2001). Gesture-space and culture. In Christian Cavé, Isabelle Guaitella, & Serge Santi (Eds.), *Oralité et Gestualité: Interactions et comportements multimodaux dans la communication* (pp. 565–571). Montréal, Paris: L'Harmattan.

Müller, Cornelia (2003). On the gestural creation of narrative structure: A case study of a story told in a conversation. In Isabella Poggi, Monica Rector, & Nadine Trigo (Eds.), *Gestures: Meaning and use* (pp. 259–265). Porto: Universidade Fernando Pessoa.

Müller, Cornelia (2004a). The Palm-Up-Open-Hand. A case of a gesture family? In Cornelia Müller & Roland Posner (Eds.), *The semantics and pragmatics of everyday gestures. The Berlin conference* (pp. 233–256). Berlin: Weidler Buchverlag.

Müller, Cornelia (2004b). Metaphors. Dead and alive, sleeping and waking. A cognitive approach to metaphors in language use. Professorial dissertation; Berlin: Free University.

Müller, Cornelia (to appear). Gesture and speech cross-culturally. Dimensions of cultural variation.

Müller, Cornelia & Harald Haferland (1997). Gefesselte Hände. Zur Semiose performativer Gesten. *Mitteilungen des Germanistenverbandes, 3*, 29–53.

Müller, Cornelia & Roland Posner (Eds.) (2004). *The semantics and pragmatics of everyday gestures. The Berlin conference*. Berlin: Weidler Verlag.

Müller, Cornelia & Gerald Speckmann (2002). Gestos con una valoración negativa en la conversación cubana [Gestures with a negative connotation in Cuban conversation]. *DeSignis, 3*, 91–103.

Müller, Wolfgang G. & Olga Fischer (Eds.) (2003). *From sign to signing. Iconicity in language and literature (Vol. 3)*. Amsterdam & Philadelphia: John Benjamins.

Nänny, Max & Olga Fischer (Eds.) (1999). *Form miming meaning: Iconicity in language & literature (Vol. 1)*. Amsterdam & Philadelphia: John Benjamins.

Neumann, Ragnhild (2004). The conventionalization of the Ring-gesture in German discourse. In Cornelia Müller & Roland Posner (Eds.), *The semantics and pragmatics of everyday gestures. The Berlin conference* (pp. 217–224). Berlin: Weidler Buchverlag.

Özyürek, Asli, Sotaro Kita, Shanley Allen, Reyhan Furman & Amanda Brown (2005). How does linguistic framing of events influence co-speech gestures?. *Gesture, 5*, 219–240. (This volume)

Pika, Simone, Katja Liebal, & Michael Tomasello (2003). Gestural communication in young gorillas (*Gorilla gorilla*): Gestural repertoire and use. *American Journal of Primatology, 60* (3), 95–111.

Pika, Simone, Katja Liebal, & Michael Tomasello (2005). The gestural repertoire of bonobos (*Pan paniscus*): Flexibility and use. *American Journal of Primatology, 65*, 39–61.

Pika, Simone, Katja Liebal, Josep Call & Michael Tomasello (2005). Gestural communication of apes. *Gesture, 5*, 41–56. (This volume)

Pizzuto, Elena & Micaela Capobianco The link (and differences) between deixis and symbols in children's early gestural-vocal system. *Gesture, 5*, 179–199. (This volume)

Posner, Roland (1994). Zur Genese von Kommunikation — Semiotische Grundlagen. In K. F. Wessel & F. Naumann (Eds.), *Kommunikation und Humanontogenese* (pp. 384–429). Bielefeld: Kleine.

Posner, Roland (2003). Everyday gestures as a result of ritualization. In Monica Rector, Isabella Poggi, & Nadine Trigo (Eds.), *Gestures: Meaning and use* (pp. 217–229). Porto: Fernando Pessoa Univ. Press.

Rizzolatti, Giacomo & Michael A. Arbib (1998). Language within our grasp. *Trends in Neuroscience, 21*, 188–194.

Roy, Alice C. & Michael A. Arbib (2005). The syntactic motor system. *Gesture, 5*, 7–37. (This volume)

Sacks, Harvey, Emanuel A. Schegloff, & Gail Jefferson (1974). A simplest systematics for the organization of turn-taking in conversation. *Language 50*(4), 696–735.

Schegloff, Emanuel A. (1984). On some gestures relation to talk. In Maxwell Atkinson & John Heritage (Eds.), *Structures of social action. Studies in conversation analysis* (pp. 266–296). Cambridge: Cambridge University Press.
Seyfeddinipur, Mandana (2004). Meta-discursive gestures from Iran: Some uses of the Pistol-hand. In Cornelia Müller & Roland Posner (Eds.), *The semantics and pragmatics of everyday gestures. The Berlin conference* (pp. 205–216). Berlin: Weidler Buchverlag.
Sherzer, Joel (1973). Verbal and nonverbal deixis: The pointed lip gestue among the San Blas Cuna. *Language in Society, 2*, 117–131.
Slobin, Dan I. (1987). Thinking for speaking. *Proceedings of the Thirteenth Annual Meeting of the Berkeley Linguistics Society*, 435–445.
Slobin, Dan I. (1991). Learning to think for speaking: Native language, cognition, and rhetorical style. *Pragmatics, 1*, 1–25.
Slobin, Dan I. (1996). From "thought and language" to "thinking for speaking". In John J. Gumperz & Stephen C. Levinson (Eds.), *Rethinking linguistic relativity* (pp. 70–96). Cambridge: Cambridge University Press.
Streeck, Jürgen (1993). Gesture as communication 1: Its coordination with gaze and speech. *Communication Monographs, 60* (4), 275–299.
Streeck, Jürgen (1994). Gesture as communication 2: The audience as co-author. *Research on Language and Social Interaction, 3*, 239–276.
Streeck, Jürgen (2002). A body and its gestures. *Gesture, 2*(1), 19–44.
Tannen, Deborah & Muriel Saville-Troike (Eds.) (1985). *Perspectives on silence.* Norwood N.J.; Ablex Publishing Corporation.
Tanner, Joanne E. (2004). Gestural phrases and gestural exchanges by a pair of zoo-living lowland gorillas. *Gesture, 4*(1), 1–24.
Tanner, Joanne E. & Richard W. Byrne (1996). Representation of action through iconic gesture in a captive lowland gorilla. *Current Anthropology, 37*, 162–173.
Taub, Sarah F. (2001). *Language from the body. Iconicity and metaphor in American Sign Language.* Cambridge: Cambridge University Press.
Terrace, Herbert S., Laura Pettito, Richard J. Sanders, & Thomas G. Bever (1979a). Can an ape create a sentence? *Science, 206*, 891–902.
Terrace, Herbert S. (1979b). *Nim: A chimpanzee who learned sign language.* New York: Alfred A. Knopf.
Tomasello, Michael & Josep Call (1997). *Primate cognition.* New York: Oxford University Press.
Tomasello, Michael & Luigia Camaioni (1997). A comparison of the gestural communication of apes and human infants. *Human Development, 40*, 7–24.
Tomasello, Michael, Barbara L. George, Ann C. Kruger, Michael J. Farrar, & Andrea Evans (1985). The development of gestural communication in young chimpanzees. *Journal of Human Evolution, 14*, 175–186.
Tomasello, Michael, Deborah Gust, & Thomas Frost (1989). A longitudinal investigation of gestural communication in young chimpanzees. *Primates, 30*, 35–50.
Tomasello, Michael, Josep Call, Katherine Nagell, Raquel Olguin, & Malinda Carpenter (1994). The learning and the use of gestural signals by young chimpanzees: A trans-generational study. *Primates, 35*(2), 137–154.
Tomasello, Michael, Josep Call, Jennifer Warren, Thomas Frost, Malinda Carpenter, & Katherine Nagell (1997). The ontogeny of chimpanzee gestural signals: A comparison across groups and generations. *Evolution of Communication, 1*, 223–253.

Tomasello, Michael, Malinda Carpenter, Josep Call, Tanya Behne, & Henrike Moll (2005). Understanding and sharing intentions: The origins of cultural cognition. *Behavioral and Brain Sciences*, *28*, 675–691.

van Hooff, Jan A.R.A.M. (1973). A structural analysis of the social behaviour of a semi-captive group of chimpanzees. In Mario von Cranach & Ian Vine (Eds.), *Social communication and movement. Studies of interaction and expression in man and chimpanzee* (pp. 75–162). London & New York: Academic Press.

von Cranach, Mario & Ian Vine (Eds.) (1973). *Social communication and movement: Studies of interaction and expression in man and chimpanzee*. London: Academic Press.

Volterra, Virginia & Carol J. Erting (Eds.) (1990). *From gesture to language in hearing and deaf children*. Berlin / New York: Springer Verlag. (1994 — 2nd Edition Washington, D.C.: Gallaudet University Press.)

Wilkins, David P. (2006). Review of Adam Kendon (2004). *Gesture: Visible action as utterance*. Cambridge: Cambridge University Press. *Gesture*, 6(1).

Wundt, Wilhelm (1921). *Völkerpsychologie. Eine Untersuchung der Entwicklungsgesetze von Sprache, Mythus und Sitte* (Vol. 1.; 4th Ed.). Stuttgart: Kröner.

BOOK REVIEW

Michael C. Corballis (2002). *From hand to mouth. The origins of language*. Princeton, Oxford: Princeton University Press.

Reviewed by Mary Copple

Introduction

The highly speculative nature of the question of how and when our hominin ancestors began to talk is cautionary, and some of the proposed answers are legendary. The truth is — whatever stories we come up with, they will always hover between fact and fiction. At best, we can only test their plausibility in view of the current state of play in disciplines that have a stake in fathoming what our language capacity is and how it works. So what evidence is currently available to us that would suggest a visuo-manual rather than an audio-vocal origin of language? Michael C. Corballis is a psychologist and an authority on brain lateralization, handedness, and visual perception. He argues in *From Hand to Mouth* for the gestural origin of language by adeptly drawing upon a vast amount of research in flourishing disciplines, such as genetics, computer science, primatology and paleoanthropology, as well as in his own specialist field of cognitive neuroscience. In addressing core issues concerning language origin and gesture, he reviews the theories and empirical studies of key authors, past and present. The broad scope of this survey will be of particular value to readers who, like myself, want to keep to pace with the debate, but who may feel daunted by the variety of rapidly developing fields involved in it. Where one's own in-depth knowledge is lacking, Corballis provides insight and orientation. Between the lines of this useful compendium of background material he interweaves his own hypothesis that, he acknowledges, has its roots in the work of the anthropologist Gordon W. Hewes who first placed the gestural theory of the origin of language on the modern agenda in the 1970s.

The story

The story that Corballis comes up with goes like this: 5–6 million years ago the common primate ancestors of humans, bonobos, and chimpanzees (our nearest biological relatives) lived in Africa. From this species two evolutionary branches diverged: the common ancestor of bonobos and chimpanzees remained quadrupedal, while hominins became bipedal. Walking enabled the first hominins to enhance their communication by enabling them to make voluntary movements of the hands. This *protolanguage* comprised a lexicon of iconic and deictic gestures as well as facial expressions. At this stage, the lexicon was not supplemented by a grammar which determined how gestures were combined, but it served as a platform upon which the language faculty was subsequently built. About 2 million years ago the genus *Homo* evolved. It included various species of hominins with large brains, who systematically made stone tools and possessed truly grammatical language in the form of a signed language. This was "punctuated with grunts and other vocal cries that were at first largely involuntary and emotional" (p. 184). Through usage, gestures became increasingly abstract and conventionalized, making communication more efficient because conventional, arbitrary symbols tend to be shorter and less ambiguous than iconic ones. By about 170,000 years ago "language was probably an amalgam of gesture and vocalization" (p. 199) produced by biologically modern humans, i.e. *Homo sapiens* with vocal tracts and brains fully capable of producing articulated speech. Following in the footsteps of their *Homo* predecessors, such as the Neanderthals, groups of *Homo sapiens* left Africa in waves of migration. In spite of their biological capacity for speech, their language nevertheless remained essentially gestural. It was handed down in tandem with their stone-tool technology until about 50,000 years ago — a pivotal date because evidence of an "evolutionary explosion" (p. 195) of sophisticated technology and art abruptly appears in the archeological record in Europe and Russia around 40–50,000 years ago.

What had happened? In a small population of about 10,000 *Homo sapiens* living in Africa prior to 50,000 years ago, language had begun to proliferate thanks to a gradual shift from the visuo-manual to the audio-vocal channel for communicational purposes. Their grunts became more articulated and intentionally controlled, while the silent gestures of their mouths, lips and tongues became voiced to expand their vocal repertoire. Speech was 'invented' as the mouth freed the hands from linguistic activity, enabling our manual dexterity to find new expression in shaping the material world. Armed with the technological advances that came with autonomous speech, members of *Homo sapiens* who could talk spread out from Africa and wiped out the gesturing indigenous hominin populations that stood in their path. Vocal autonomy promoted cultural innovation to the extent that today "culture has replaced biology as the main source of human accomplishment and variation" (p. 219). The fact that we all gesticulate as we talk, even on the telephone, is viewed by Corballis as living evidence that gestures were at the evolutionary root of our linguistic capacity — gesticulation is but a vestige of their former primacy.

The evidence

The sequence of events in this story is not new. Corballis admits that his account of how speech took over from gesture is "scarcely any improvement" (p. 64) on Condillac's (1746) scenario. And, more recently, a similar path unfolds in Leroi-Gourhan's (1964) reading of the paleoanthropological and archeological records. What is new is matching the transition from gestural to vocal language with stages in hominin evolution along a timeline that would generally meet with consensus among researchers today. So what kind of evidence, other than our vestigial gesticulations, does Corballis call upon to support his story? The fossil record of our ancestry is extremely sparse and subject to reinterpretation as new finds deliver new pieces to be integrated into the evolutionary puzzle. Moreover, the definition of 'species' is a hot topic that bears on the classification of both extinct and extant life forms. In addition to traditional criteria, such as distinctive morphology and whether two populations can produce fertile offspring, genetic lineages have come into play as a new measure of evolutionary divergence.[1] Corballis skillfully guides the reader through the expanding museum of hominin bones with a contemporary map of speciation, indicating cultural milestones of humanity, such as stone tools and cave art, along the way. He discusses our distinctive bipedalism not just in relation to the possibilities of enhanced gestural communication that it opened up. He also explores controversial theories concerning the range of environments — savanna, woodland, riverbanks, lakesides and coastal regions — with which the remains of hominins are associated, and how changes in habitat may have created the selective evolutionary pressures that promoted upright stance.

The genetic evidence that underpins Corballis' timeline along which talking *Homo sapiens* emerged involves juggling with distant dates with generous margins. This evidence is based on the assumed uniform rate of mutation and degree of variation in the Y chromosome in the male line, and the degree of variation of mitochondrial DNA (mtDNA), which is generally considered to be only passed down the female line.[2] These factors are used to estimate how much time has elapsed since the most recent common ancestor of the people who contributed to the analysis was alive. The estimates to which Corballis refers divide the current world population into Africans and non-Africans. One estimate using mtDNA suggests that "the nearest common ancestor of both African and non-African people lived as recently as 52,000 years ago [plus or minus 27,500 years], although the African lineages themselves go deeper, to some 170,000 years ago" (p. 134), plus or minus 50,000 years. Other estimates, using mtDNA and Y-chromosomal data, would generally support the latter period for an ancestral Adam and Eve living in Africa.

On the basis of these data Corballis selects a time window of 50–170,000 years ago for the modality shift from gestural to vocal language. In keeping with Philip Lieberman's (1998) theory that the physical prerequisites for the production of articulated speech (essentially a lowered larynx) were in place around 150,000 years ago, he sets the base line for the beginning of this gradual shift at 170,000 years ago. If the vocal equipment for producing speech was in place around 150,000 years ago, why then does Corballis opt for a date closer to 50,000 years ago for vocal autonomy? Firstly, because a wave of migration out of Africa around that time would be consistent with estimates of when the most recent common ancestor of all present-day non-Africans and Africans was alive. And secondly, because such a migration of fully-fledged talkers would neatly coincide with evidence of the "evolutionary explosion" (p. 195) of culture that occurred in its wake — and which could be explained as a consequence of the advantages that speech conferred. A direct consequence, Corballis suggests, was the superior technology that gave the talkers the edge over the indigenous gesturing hominin populations they replaced. He sees cave drawings as an "indirect consequence" (p. 201) of vocal autonomy and, like Leroi-Gourhan, he links this representative activity of the hand to the subsequent development of its new linguistic function: writing. In sum, these consequences may have produced evidence of the modality shift that he proposes, but he does not suggest what may have caused it — what prompted our ancestors to invent autonomous speech at that time.

Corballis argues for his version of prehistoric events by discussing parallels, differences, and interactions between forms of animal and human communication that are observable today. Gestures, vocalizations, toolmaking activities, as well as physical mobility and cognitive processing in general are all taken into account. Such behavioral studies and experiments raise the question about the extent to which comparative behavioral research can inform us about human origins and especially about our language capacity now. The similarities between birdsong and speech provide an example of "convergent evolution — independent adaptations to common environmental challenges" (p. 3). But his main focus is on nonhuman primates in captivity and in the wild. As he recounts, our nearest biological relatives can, admittedly with great effort on their part and that of their trainers, learn a kind of protolanguage by communicating with us through gestures or keyboards of abstract symbols, and to some extent even 'understand' human speech. This is remarkable. But their capacity to produce anything that we would recognize as languagelike seems to be far more limited. If rules do underlie their novel sequences of two to three gestures or symbols, these do not approach anything like the complexity of grammar and, as all Chomskyans would promptly point out, they crucially lack recursion. Corballis is sceptical about some of the claims made about the linguistic abilities of domesticated nonhuman primates. Nevertheless, he uses the fact that we can communicate with them so much better via the visuo-manual than the audio-vocal channel to infer that our common ancestor "would have been much better equipped to develop a communication system based on manual and bodily gestures than one based on vocalization" (p. 32). Regarding his catalogue of gestures that great apes are reported to make in the wild, it is tempting to consider these as evidence of a potentially shared basis for primeval symbolic activities. But it could equally well be just a compilation of researchers' interpretations aimed at building bridges between their own ways of communicating and those of the animals they are investigating. This is even more so the case in studies involving animals in captivity: there is a danger that mutually conditioning dialogues between the researching and the researched creatures can subjectivise the interpretation of data. Furthermore, captive apes that take part in experiments to assess their symbolic capacity may also reapply what previous experiments have trained them to do. But researchers are aware of these dangers. So, as well as experimenting, they also use their trained eyes to discretely observe the spontaneous communicational interactions between captive animals. Moreover, the increasing use and rapidly improving quality of digitized audio-visual data should enable greater objectivity and accelerate progress towards gaining consensus on theories of gestural communication among nonhuman primates. For the time being, Corballis concludes that "great apes gesture extensively in the wild" (p. 215), and it appears that they gesture 'naturally' in captivity too. But whatever we judge their hands and faces to be *saying*, how can we ever know what is really going on inside their heads?

Regarding the neurological aspect of this question, technology may provide some insight in the form of brain-imaging studies: positron emission tomography (PET) and magnetocephalography (MEG) are used to show which areas of the brain are engaged in specific activities. Corballis is an authority on brain lateralization and handedness, as his chapter on cerebral and manual asymmetries in humans and nonhuman species admirably demonstrates. He not only draws upon his own research to link language and gesture neurologically, but he also brings in the evidence of *mirror neurons* in the brains of both humans and apes that has been produced by Giacomo Rizzolatti, Luciano Fadiga, Vittorio Gallese, and Leonardo Fogassi (1996). Their PET and MEG scans show that particular neurons respond when a monkey makes a particular movement — and when he/she perceives the same movement being made by a human being. Hence, they "provide a mirror between action and perception. What you see is what you *do*" (p. 46). In humans, mirror neurons are thought to be located in several areas of the brain, including Broca's area in the left cerebral hemisphere that is clearly associated with the production and perception of both gestural and vocal language. In macaque monkeys, mirror neurons are claimed to be located in the areas corresponding to Broca's area on *both* sides of the brain. They are seen to be activated during grasping actions and the perception of them, but not during vocal production or perception. On the basis of this evidence, Corballis proposes that "somewhere in the progression from ape to human, vocalization and gesture were linked, and the system became lateralized. My guess is that this happened when vocalization was added to the gestural repertoire and synchronized with it" (p. 169).

What could mirror neurons have to do with language? In humans, they may underlie our ability to imitate each other's actions, and to see the world from other people's perspectives, usually referred to as *theory of mind*. Corballis stresses the visual foundation of theory of mind, which is evident in recursive thinking and which, in turn, is evident in sentences containing relative clauses: "the structure of the understanding that 'I know that he can see me' has the same recursion as the sentence that expresses it" (p. 60). As regards our imitative abilities, language acquisition clearly depends on copying sound patterns. And in young children, mirror neurons may play a role in perceiving the articulatory movements that (re)produce the sounds of words, although perhaps not in learning the arbitrary sound-meaning connections that constitute them (Hurford, 2003). Liberman and Whalen's (2000) update of the Motor Theory of speech perception lends weight to Corballis' argument that mirror neurons may have enabled our ancestors to imitate and use gestures of the hands and face in languagelike ways long before the modality shift to speech took place. Such gestures could have been made voluntarily and with greater flexibility than animal calls, which are typically involuntary responses to fixed situations. So, iconic movement-meaning connections could have been established which enabled intentional responses to novel situations to be made. In contrast to humans, mirror neurons in nonhuman primates do not apparently enable any kind of imitative learning (Tomasello, 1999). And whether or not any of our primate cousins have theory of mind is a very controversial issue. Corballis suggests that gesture incrementally shaped our capacity for recursion *after* our species split from the common ancestor we share with chimpanzees and bonobos — which is why they display only limited theory of mind, if they possess it at all, and why they do not share our capacity for language.

This leads us to the crux of the matter and into a terminological minefield: what does Corballis mean by *language*? He devotes the first chapter of his book to this very question. Linguists do not agree on what language is, and Corballis' conception of language is largely influenced by linguists who "like to draw a clear distinction between grammar and meaning" (p. 6), and especially by Noam Chomsky's theory that recursion constitutes the core of language. Corballis' own theory involves a string of other terms with which *language* inevitably becomes entwined, such as *thought, representation, idea, concept, word, sign, icon*, and, last but not least, *meaning*. He explains that "language is not just speech" (p. 16). It is also *thought*. But because thought is also indisputably evident in signed languages and because these have no basis in sound, it "runs deeper than speech" (p. 17). Thus, thinking is "not always" (p. 17) internal speech. He conceives of a *language of thought* that is independent of words: "nonverbal thinking depends on our ability to represent objects, sounds, and actions in our minds and to manipulate them mentally" (p. 17). Imagination, fantasy, and problem solving can be the products of "nonverbal, and often spatial" (p. 18) thoughts, rather than linguistic ones. The creativity that this kind of thinking can produce is exemplified by Einstein, who apparently imagined himself traveling on a beam of light as he worked out the theory of relativity. So, *mental representations* form the building blocks of Corballis' understanding of language that words or gestural signs can express. His *language of thought* is similar to Pinkerian *Mentalese* but without the total rejection of the proposal that particular languages and individuals capture particular worldviews.[3] Indeed, he readily acknowledges that ideas and concepts can vary across languages and, therefore, "not all cultures think alike" (p. 11). He asks if language could be what Richard Dawkins (1976) calls a *meme* — a culturally determined characteristic, such as stories, songs, beliefs, inventions, political systems, and cuisine. Glossing over the central role that language plays in transmitting memes such as these, Corballis asserts that, in some respects, language is a meme because "the actual words we use are passed on by the culture we live in" (p. 15) — or does he really mean its speakers/signers? He concludes that, although the learning process itself is genetically determined and goes beyond our capacity to imitate sequences of sounds or movements, language must be learnt because of its cultural aspects and the arbitrary nature of words.

Corballis' angle on the arbitrary nature of words is revealed in a chapter devoted to signed language. This begins with a compact history of the slow recognition that signed languages are truly languages although they are not based in sound. Corballis then presents theoretical maps of the structure of verbal and signed language which are brought to converge to see how they match up. So, how do their units compare — what is a *sign*? For him, a sign is a gesture in a signed language. It is "the basic unit that corresponds to the *word* in spoken languages" (p. 110). And what is a *word*? His use of this term is molded by his interpretation of the *signe linguistique* coined by Saussure (1916) whereby "the sequence of sounds that make up a spoken word, which [Saussure] called the *signifiant*, and what it represents, the *signifié*" (p. 111) are distinguished, and whose relation is arbitrary.[4] Hence, "the words we use typically bear no relation to what they represent" (p. 110). He views the "distinction" between *signifiant* and *signifié* as often "blurred in signed language, where signs have a more iconic (or pictorial) relation to what they represent", although this is "often an illusion, in that the pictorial aspect is only evident after one has discovered what the signs actually stand for" (p. 111). Interestingly, Corballis sees a sliding scale of iconicity in signed languages in which "iconicity and arbitrariness are not opposites, but rather the ends of a continuum" (p. 112), along which signs tend to progress from iconic to arbitrary as their initial novelty wears off and their frequency of usage increases. He argues that this process of conventionalization is driven by the communicational advantages of arbitrary signs. They are shorter, more efficient, and "many of the concepts we communicate about are abstract in the first place, and the signs representing them cannot be decoded from the gestures comprising them" (p. 112 f.). Furthermore, arbitrary signs are also able to avoid confusion where iconic signs cannot readily distinguish between similar actions and objects, for example, between Ford and Chevrolet cars. So, it is its greater potential to make distinctions (a central Saussurean principle) that gives arbitrariness the edge over iconicity when it comes to optimizing language, especially for communicational purposes: "A well-designed language makes use of maximum contrasts, so as to optimize the fidelity and clarity of the message" (p. 114). The apparent lack of iconicity in spoken language demonstrates this, and Corballis does not fail to point out that onomatopoeic expressions are few in number in any language and differ across languages.

The process of conventionalization in the evolution of signs which Corballis proposes retraces Condillac's theoretical progression from *signes naturels* to speech: concrete actions led to iconic signs for actions and objects (feelings and mental representations expressed through imitative sounds and gestures), then to conventional bimodal symbols (arbitrary gestures and words), which formed combinations and thus comprised a protolanguage (*langage d'action*), and whose modalities subsequently split and gave rise to compositions of silent movements (dance) and truly grammatical spoken language (articulated speech) (Condillac, 1746, p. 193 f.). One vital aspect that distinguishes their models is a 20th century linguistic insight into the (possibly unique) structure of language called *duality of patterning*: firstly, a verbal message can be broken down into minimal distinctive units of meaning (morphemes)[5]; and secondly, each morpheme can be broken down into minimal distinctive units of sound (phonemes)[6] of which there is a specific, limited range in any given language. It is phonemes which guarantee the arbitrary nature of the linguistic sign (Martinet, 1957). Hence, duality of patterning can be seen as the structural characteristic of language that provides the basis of its arbitrary nature. Corballis aims to correlate units of signed language with the units on both these levels that interlock in verbal language. He calls the grammar that determines how sounds combine to build words *phonology*, and the grammar that determines how words combine to form sentences *syntax*. Each level is discussed under these headings in order to establish if the same principles hold for signed and verbal languages.[7] This analysis is central to his evolutionary theory, and it addresses an issue that is common to all gestural theories of the origin of speech: the relation between modality shift and structural change. Corballis questions the existence of acoustic phonemes and, in support of the Motor Theory of speech perception, he considers speech as voiced gestures: "Some phonemes, at least, have little acoustic reality at all and may even be an artificial product of literacy. It may be more appropriate to think of speech, not in terms of combinations of those phantom entities called phonemes, but rather as combinations of sound 'gestures' that we can make by the deployment of six independent 'articulators' in the vocal tract." (p. 118 f.), i.e. lips, blade of the tongue, body of the tongue, root of the tongue, velum/soft palate, and larynx. So, the suggestion is that the transition from movement to sound as the medium of linguistic expression may have been facilitated by the addition of voice to previously existing silent gestures of the face and vocal tract.

Putting his scepticism about the physical existence of phonemes to one side, Corballis takes a bottom-up approach to duality of patterning in speech and deals with *Phonology* first. A phoneme materializes differently depending on its neighboring phonemes, and on the voice of the speaker pronouncing it. And although we do distinguish between phonemes, it is intriguing that some do not show up on sound spectrographs at all. How our brains create distinctions between sounds that machines cannot detect is not known. But our ability to identify constancy within variation is the touchstone of Corballis' structural comparison between speech and signed language. So, what are the gestural equivalents of acoustic phonemes? These turn out to have a very different nature because speech materializes one-dimensionally in time as linear sound sequences, whereas signed language materializes four-dimensionally in space-time as visually perceivable movements of the body. Corballis opts for the system devised by William C. Stokoe, Dorothy C. Casterline, and Carl G. Croneberg (1965) which identifies "three different kinds of components, not two as in the consonants and vowels of spoken language", namely "*tab* for location, *dez* for handshape, and *sig* for movement" (p. 117). Whereas consonants alternate with vowels in a linear sequence to form a word — tab, dez, and sig materialize simultaneously to form a sign. For example, the sign for *cheeky* in British Sign Language is composed of the cheek (tab) held between the thumb and index finger (dez) and shaken to and fro (sig). So, "the different possibilities available for each tab, dez, and sig are equivalent to phonemes and constitute the 'phonology' of signed language" (p. 117 f.). And, like phonemes, these possibilities are limited to different ranges in different signed languages, and there is variation in the materialization of each component depending on the neighboring signs, and on the person making the sign. But does the parallel with this level in duality of patterning in speech diverge? Whereas the linearity of speech enables vowels and consonants to be discerned individually as discrete units of sound (phonemes),[8] the spatio-temporal nature of signed language enables variables of tab, dez, and sig to be discerned simultaneously in discrete units of meaning (signs). So, each sign is a kinetic amalgam with three distinguishing facets or 'parameters', each of which performs the same *function* as an acoustic phoneme — but has a sign an internal *articulation* comparable to that of an acoustic morpheme? Is a sig (a hand movement) discrete in the same way that a phoneme (the product of a vocal-tract movement) is? One would expect the characteristics of the communication channel employed to impact the connection between the form and the meaning of the message, and hence the encoding of the latter.[9] One would expect different modalities of linguistic expression to reflect differences in the way they pattern/articulate thought. But perhaps the translation of the French term *double articulation*, as formulated by André Martinet (1949), into English as *duality of patterning* obscures insights into the specific nature of speech. Moreover, interesting contrasts with the more amalgamated nature of signed language may be missed at the expense of bringing out its "underlying similarity" (Knapp & Cheek 2002, p. 27) with speech in order to underline that its nature is truly, equally linguistic. Exploring differences in the way that language is articulated in these different modalities may well further our understanding of what linguistic cognition is and how it works.[10] The extent to which tab, dez, and sig equate with acoustic phonemes is up to the reader to decide, but the availability of such recomposable elements is clearly a common feature of both the visuo-manual and the audio-vocal modalities of language. Hence, Corballis' analysis of the kinetic organization of signs underpins his argument that "in signed languages, individual signs are already generative, in that they can be constructed from different combinations of the basic elements — tab, dez, and sig — and the next level of generativity, the combinations of the signs themselves, would follow from this" (p. 125).

Under the heading *Syntax*, Corballis reveals structural differences in semantic organization between signs and words in sentences. Once again, these are determined by differences in the medium of expression: the possibilities offered by simultaneous signing contrasts with the constraints imposed by the linearity of speech. For example, a shake of the head while signing in American Sign Language (ASL) negates a statement, whereas the word *not* in spoken English would be inserted in a verbal sequence. And reference to time, people and places can be made through the specific use of spaces in relation to the body rather than morpho-syntactically. However, signed language does possess some of the syntactic properties of verbal language. For example, in ASL the sign sequence subject-verb-object is the norm, just as it is in English sentences. Corballis demonstrates that a range of grammatical features of verbal language have equivalents in signed language. He concludes that "it seems that virtually every aspect of syntax as identified in spoken language finds its counterpart in ASL" (p. 122). Signs are ordered into sequences along grammatical guidelines in the interplay between space and time. Unlike the tendency of iconic signs representing objects to migrate towards abstraction and conventionalization, "this does not appear to be true of the syntactic use of space and time, which remain strongly iconic" (p. 123). For example, some signers use an imaginary time line stretching from behind (past) to in front of them (future) to temporally situate events.

Corballis rounds off his comparison of how signs and words combine to convey meaning with a top-down perspective which reveals a vital difference between them: many signs are holophrastic, i.e. they are equivalent to whole sentences (e.g. ex-prime minister Pierre Trudeau's characteristic shrug meaning *who knows*?), whereas few words can be used in this way (e.g. *sorry*!). This point is a major pillar of his evolutionary story. It has its foundation in the theory of David F. Armstrong, William C. Stokoe, and Sherman E. Wilcox (1995) that "in signed language, individual signs have the same basic structure as a sentence" and hence, "the structure of the sentence is derived from the sign itself; [...] the sign is the *seed* for syntax" (p. 123 f.). Although I have no doubt that the various pieces of information that can be integrated into a single gesture can be fully transformed into a grammatical sequence of words, and vice versa, I do not see how the *structure* of a holophrastic sign could be a template for its verbal equivalent because it has by definition no duality of patterning — no internal structure at all apart from its phonological components: the whole gesture is the whole message. Nevertheless, Corballis adopts the theory of Armstrong et al. (1995) and suggests that, during the course of human evolution, our ancestors began to combine holophrastic signs with signs that were equivalent to words representing objects and actions, and that this gave rise to our capacity for recursive grammar.

In this, Corballis steers a course between two polarized positions in the current debate about language evolution: the holistic (or analytic) vs. the synthetic approach to accounting for speech. The former is more in keeping with Corballis' view, and indeed does not exclude holistic gestures (holophrastic signs) from playing a role in the formative stages of language evolution. For example, Alison Wray (2000) proposes that speech had its roots in utterances comparable to the holistic noise/gesture signals made by chimpanzees. She argues that, since people often employ holistic utterances in the form of formulaic sequences (e.g. *happy birthday*) and, in common with our primate cousins, use them to fulfill prime social functions (e.g. social integration and interpersonal manipulation), holistic language existed for the purpose of communication before and after acoustic protolanguage emerged. Hence, the continuity afforded by holistic language would have provided a stable framework for the cognitive enhancement that the advent of grammatical language entailed: while the former ensured successful interpersonal communication, the initial function of the latter was to create a private space for "talking to oneself" (Wray, 2000, p. 291). Thus, regardless of survival pressures, grammatical language was able to generate the benefits of organizing creative thought and planning by producing word-sized concepts, referentiality and hierarchic structures that juxtaposed ideas in different ways. Wray suggests that holistic utterances provided the material for this development by undergoing two transitions that need not have occurred simultaneously: from mimetic/iconic to arbitrary representation and from nonphonetic to phonetic materialization. Thus, holistic utterances were unanalyzed units that became segmented into meaningful subunits — a process triggered by a phonetic unit (e.g. a syllable) occurring by chance in two or more holistic units that contained a common semantic element. Hence, an arbitrary sound-meaning connection became established. She proposes that the process of recognizing and correlating phonetic/semantic patterns, and hence forming segments, became increasingly rationalized and gradually gave rise to full compositionality. Unlike Corballis' story, in Wray's scenario holistic gestures fade into oblivion as the segmental processing of their acoustic counterparts kicks in: it is the temporal, linear nature of holistic sound sequences that promoted segmentation into meaningful subunits at syllabic boundaries — the fault lines along which they became detached from their phonetic neighbors. Perhaps because of their four-dimensional nature, and in particular because of the fluidity of their movement component, holistic gestures would have been inherently less apt to fracture and more difficult to analyze into discrete units?[11]

In contrast to the holistic theory, the synthetic approach to accounting for the evolution of speech posits that protolanguage was based on single words which came to be combined randomly at first and later formed structured sequences as syntax emerged. Many of the objections to holistic scenarios are raised by Maggie Tallerman (2007), who claims that these tend to ignore crucial factors. Among those concerning structural issues are some which are pertinent to scenarios rooted in holistic gestures: "holistic accounts [of the evolution of spoken language] have simply assumed the prior existence of modern phonetic segments" (Tallerman, 2007, p. 586), i.e. consonants and vowels, instead of accounting for their emergence. And they have assumed that no phonetic change occurred over many thousands of years, because without phonetic stability the analysis and segmentation of holistic utterances would not have been possible. Tallerman counters that no phonological system is static: all languages are subject to principles of phonological change over time (Tallerman, 2004, p. 8). And with respect to duality of patterning, it is assumed that a phonological system predated segmentation: she objects that "morphemes can never be extracted from holistic utterances", because "you can't have morphemes without phonemes (since morphemes are composed of phonemes) and you can't have phonemes without words, since you have to have semantic contrasts and minimal/near-minimal pairs in order to know what the phonemes are." (Tallerman, 2007, p. 587 f.) Otherwise our ancestors would not have known which phonetic distinctions were to be treated as significant. Assuming that spoken and signed language are both subject to principles of phonological change and those underlying duality of patterning, these points of contention also have to be addressed by scenarios involving the breakdown of holistic gestures into the gestures which make up a fully-fledged linguistic system. Corballis' suggestion that our ancestors combined holophrastic signs (holistic gestures) with referential signs that were equivalent to words could indicate a middle way that mediates between these polarized positions.

Had Corballis taken the original approach to tackling duality of patterning and begun with a unit of sense — a message — breaking down into minimal distinctive units of meaning (Martinet, 1949, 1957), then his analysis may have been further enriched by tapping the resources that the study of *Semantics* opens up. This goes beyond inquiry into the lexicon. Considering how the meaning of a message is distributed across its constituent morphemes includes considering their syntactic relations and could reveal structural overlaps and differences between the verbal and the gestural forms that deliver it. In particular, the role that tropes, such as metaphor, play in encoding world-views in linguistic forms is an inquiry conducted within cognitive semantics that may be relevant here: metaphors, for example, are a major source of linguistic innovation that are not only a vital element in poetry but also pervade everyday speech with novel uses of language.[12] Is there a link between metaphoricity in spoken/written language and iconicity in signed language? Once again, highlighting the ways in which people interweave semantic fields differently in both modalities, as well as the common ground between them, may be enlightening.

Springing from his evolutionary perspective on language structure back to the present, where we can observe modern human babies developing their language faculties, Corballis asks the classic question: what is due to genetics and what to learning through experience? Words, he answers, must be learned because of their arbitrary nature. Can grammar be inductively learned — or — does innate universal grammar become parametized through exposure to linguistic data? As an alternative to the latter option — the Chomskyan computational model of language acquisition — Corballis favors the connectionist view based on a model of the brain as an associative device capable of learning recursive grammatical rules simply through experiencing language. The evidence supporting his position is provided by the neural networks of Jeffrey Elman, Elizabeth Bates and Elissa Newport (1996) that *grow*, i.e. initially process "only global properties of the input" (p. 14) and then become more finely tuned as they feed on words. Their research demonstrates that rules can emerge out of associations although, decisively, these have yet to attain the level of recursion. The suggestion is that grammar does not have to be innate because such networks may be capable of mirroring very basic grammar learning in children. Technological breakthroughs such as these are used to bolster the argument that language could be an integral part of general intelligence as opposed to operating within a separate, genetically specified *language acquisition device* as tenacious Chomskyans maintain. The "*generative assembling device*" (p. 206) that Corballis favors instead would include language in a range of skills that display recursion, and whose common cognitive core differentiates according to purpose, such as creating music. Indeed, he talks about talented young musicians "impregnated with the *grammar* of music" (p. 206).

In his view, the software that Elman et al. (1996) have created may fall short of being able to account for "many of the niceties of grammar and meaning", but they keep language within "the realm of biology" (p. 15), where he believes it belongs. As an alternative to Pinkerian grammar genes, Corballis proposes that "biologically programmed patterns of growth" (p. 206) plus experience may determine the biological differentiation of neural tissue in and around Broca's area. His evidence? There is a phase of growth in the left side of the brain in two to four-year olds as they are acquiring skills in language, object manipulation and theory of mind. Thus, this part of the brain, traditionally associated with language, may support a range of activities that all have hierarchical representations of the world and recursion in common. He theorizes that patterns of growth and patterns of representation may knit together to develop grammatical and other generative abilities that apply the same basic principles. He argues that this could explain "why grammar is amodal, as much at home in gesture as in speech", and that their "details of structure" (p. 208) may depend on other constraints imposed by the medium of expression. Placing this evidence in an evolutionary perspective, Corballis thinks that "our unique capacity for language may depend simply on evolutionary alterations to the growth pattern" (p. 14), i.e. extended postnatal growth, larger brain/body ratio and changes in the relative sizes of brain parts. Thus, the genetic foundation of the language faculty may boil down to "modifications that have altered the basic body plan of animals throughout biological evolution" (p. 14).

Conclusion

Corballis weighs up a vast amount of data from a variety of disciplines in a well-structured text with lots of clear examples that make the result accessible to a wide readership. His conclusion? His "guess" is "that gestural languages, then as now, lie closer in evolutionary terms to the biological adaptations that gave rise to grammar and the ability to represent objects and actions internally, and depend less on cultural input. In a word, signed language may be more *natural*" (p. 108 f.). In a sentence, gesture is a natural adaptation, and speech is a cultural invention. In arguing his case, he achieves the unlikely and admirable feat of being not only scholarly but also humorous. This delightfully engages the reader in often complex stories emerging at the frontiers of current research. These may not be entirely new, but the evidence upon which their plausibility depends certainly is.

On the one hand, Corballis considers real bodies and real brains in which cognition and communication are truly woven together in real people in the real world. On the other, he affiliates to endeavors to create computer simulations of brains that deliver maps of minds — a real quest that echoes the goal of neurophenomenology revealed in Dan Lloyd's novel, *Radiant Cool*, "to find a transparent theory of consciousness, a Rosetta Stone" in which "you'd put in phenomenology at one end and get spiking neurons at the other" (Lloyd, 2004, p. 31). Can we imagine such a Rosetta Stone without words, or their gestural equivalents, in the middle? How does human consciousness hang together with the (neuro)nets of meaning spun by our linguistic thoughts? The interface between the brain and the mind has yet to be mapped. How bioelectricity flowing through neurons translates into thought is still a mystery. Corballis' book is a valuable contribution to consolidating the central position that gestural studies occupy both historically and currently within the framework of this exciting and challenging debate.

Notes

1. Given that there can be both morphological diversity within a species and morphological similarities between species, it can be difficult to reach a consensus on which species a hominin fossil, sometimes just a fragment of bone, belonged to. Tattersall (2001) addresses taxonomic issues concerning the human fossil record, criticizing the classification of claimants to the status of early *Homo* on the basis of their potential association with stone tools where morphological evidence is inconclusive. Pairing up genetic lineages with species introduces a new dimension to such controversies. The recent discovery of the remains of the smallest hominin species known, *Homo floresiensis*, on the island of Flores in Indonesia (Brown et al., 2004; Morwood et al., 2004) reveals the morphological diversity of the genus *Homo*. The most complete skeleton is that of an adult female displaying characteristics which correspond to different stages in hominin

evolution. With her tiny human-like hands and the smallest-known hominin brain (endocranial volume: 380 cc) relative to body height (stature estimate: 106 cm), she belonged to a species that evidently used fire to cook its prey and made stone tools, although it is widely considered that such capabilities correlate with a much bigger brain/body ratio in the hominin fossil record. The authors claim that this species lived before 38,000 and up to 18,000 years ago — a period which overlapped the presence of *Homo sapiens* in the region — and that it originated from an earlier migration of *Homo erectus* out of Africa. This would concur with Corballis' position that various hominin species coexisted and competed at particular points in time.

2. Philip Awadalla, Adam Eyre-Walker, and John Maynard Smith (1999) argue that mitochondrial DNA can also pass down the male line.

3. Cf. Trabant's (1996) historical perspective on the contemporary study of linguistics that excludes semantics from its consideration and his critique of Pinker in particular.

4. Corballis' interpretation of Saussure's (1916) *signe linguistique* is somewhat ambiguous. Just in case the reader is led to make the common mistake of confusing the *signifiant* with the spoken word, and the *signifié* with the thing the word is referring to (*referent*), it is important to explicate Saussure's linguistic conception here. This is because Saussure clearly refutes the Aristotelian position that words are arbitrary names for things, and that our concepts of them are universally the same and language independent: "Le signe linguistique unit non une chose et un nom, mais un concept et une image acoustique" (Saussure, 1916, p. 98), i.e. "the linguistic sign does not unite a thing and a name, but a concept and an acoustic image" (my translation). The *signifiant* is the acoustic image that a spoken word imprints in the mind, not the physical sound waves that create the acoustic image. The *signifié* is the concept that a *signifiant* evokes. So, when someone speaks, a selection of his/her stored *signes linguistiques* form sequences that materialize. The acoustic images thus imprinted on the listener's mind, together with the concepts they evoke, must be similar enough to those of the speaker in order to understand what he/she is saying. The concept on the semantic side and the acoustic image on the phonetic side of the *signe linguistique* are associated in an inseparable bond — like the two sides of a piece of paper — and both sides are language specific within a historical tradition. The *signe linguistique* can be said to be arbitrary in that the composition of both the *signifiant* and the *signifié* is culturally determined; it does not manifest a natural link to what the *signe linguistique* is used to refer to. Hence, a language is a system of distinctive units of thought-sound (*pensée-son*) that articulates the world-view of the speakers/listeners who inherit, maintain and modify it according to their needs (cf. Trabant, 1996). Corballis' interpretation of the Saussurean *signe linguistique* may miss these insights but not the relevance of its arbitrary nature.

5. Some morphemes can stand alone as well as join together in polymorphemic words, e.g. *black* and *bird* in *blackbird*. Others cannot stand alone, e.g. affixes such as *un* and *ish* in *un*self*ish*, and the *s* in blackbird*s* that expresses plurality (cf. Crystal, 1997, p. 12 & p. 248).

6. For example, the /b/ sound in *bat* that distinguishes it from *cat*.

7. Some linguists may object to this terminological usage on etymological grounds: 'phono' stems from the Greek word for 'sound'. Hence, it may seem inappropriate to talk about the 'phonology' of signed languages. However, this term was coined by linguists inquiring into speech, and it seems appropriate to reapply it to signed language because the discrete minimal units of meaning in both modalities can be further decomposed into discrete minimal distinctive units.

Regardless of how these may differ in their material realization, they share the same function. Therefore, accepting a common term for and exploring this common level of analysis in spoken and signed languages could reveal essential differences that further our understanding of the interplay between linguistic structure and its materialization.

8. The phonemes that make up a morpheme in speech are not completely discrete: "the acoustic cues that convey the individual sounds of speech that roughly correspond to the letters of the alphabet are melded together" (Lieberman, 1998, p. 15) in that consonant and vowel sounds spread across syllables during articulation (cf. Brentari, 2002, p. 45.)

9. Meier, Cormier, and Quinto-Pozos (2002) offer a range of research papers which examine how linguistic structure may be affected by the mode in which language is produced and perceived.

10. Diane Brentari's (2002) comparative analysis of phonology in ASL and spoken English is a prime example of this kind of research.

11. Recent research into Nicaraguan Sign Language (NSL) has produced empirical evidence that the linguistic properties of discreteness and combinatorial patterning can emerge in the visuo-manual modality alone. Since the early 1980s when Nicaraguan deaf children first invented NSL, successive generations of new learners/producers have continued to develop it. Senghas, Kita, and Özyürek (2004) claim that the linear sequencing of semantic elements in NSL, e.g. expressing the manner and path of a motion event such as rolling down a hill, is tending to override previously simultaneous combinations. They assert that this may be explained by two learning mechanisms available to children: "The first is a dissecting, segmental approach to bundles of information; this analytic approach appears to override other patterns of organization in the input, to the point of breaking apart previously unanalysed wholes. The second is a predisposition for linear sequencing; sequential combinations even appear when it is physically possible to combine elements simultaneously, and despite the availability of a simultaneous model" (Senghas et al., 2004, p. 1781).

12. Jean Aitchison (1994) quotes a survey in which "there were, on average, over five examples of figurative language per 100 words spoken, almost a third of which were novel uses" (p. 149).

References

Aitchison, Jean (1994). *Words in the mind: An introduction to the mental lexicon* (Second edition). Oxford and Cambridge, MA: Blackwell Publishers Inc.

Armstrong, David F., William C. Stokoe, & Sherman E. Wilcox (1995). *Gesture and the nature of language*. Cambridge: Cambridge University Press.

Awadalla, Philip, Adam Eyre-Walker, & John Maynard Smith (1999). Linkage disequilibrium and recombination in hominid mitochondrial DNA. *Science* 286, 2524–2525.

Brentari, Diane (2002). Modality differences in sign language phonology and morphophonemics. In Richard P. Meier, Kearsy Cormier, & David Quinto-Pozos (Eds.), *Modality and structure in signed and spoken languages* (pp. 35–64). Cambridge: Cambridge University Press.

Brown, P., T. Sutikna, M.J. Morwood, R.P. Soejono, Jatmiko, E. Wayhu Saptomo & Rokus Awe Due (2004). A new small-bodied hominin from the Late Pleistocene of Flores, Indonesia. *Nature*, 431, 1055–1061.

Condillac, Etienne Bonnot de (1746). *Essai sur l'origine des connoissances humaines, ouvrage où l'on réduit à un seul principe tout ce qui concerne l'entendement* (ed. Charles Porset). Auvers-sur-Oise: Galilée 1973.

Crystal, David (1995). *The Cambridge encyclopedia of language*. Cambridge: Cambridge University Press.

Crystal, David (1997). *A dictionary of linguistics and phonetics*. Fourth edition. Oxford and Cambridge, MA: Blackwell Publishers.

Dawkins, Richard (1976). *The selfish gene*. Oxford: Oxford University Press.

Elman, Jeffrey, Elizabeth Bates, & Elissa Newport (1996). *Rethinking innateness: A connectionist perspective on development*. Cambridge, MA: MIT Press.

Hurford, James R. (2003). Language beyond our grasp: What mirror neurons can, and cannot do for language evolution. In Kimbrough Iller, Ulrike Griebel, & Kim Plunkett (Eds.), *The evolution of communications systems: A comparative approach*. Cambridge, MA: MIT Press.

Knapp, Heather & Adrienne Cheek (2002). Phonological structure in signed languages. In Richard P. Meier, Kearsy Cormier, & David Quinto-Pozos (Eds.), *Modality and structure in signed and spoken languages* (pp. 27–33). Cambridge: Cambridge University Press.

Leroi-Gourhan, André (1964). *Le Geste et la Parole*: Vol. 1: *Technique et Langage*. Vol. 2: *La Mémoire et les Rythmes*. Paris: Éditions Albin Michel.
English translation (1993). *Gesture and speech* (tr. Anna Bostock Berger). Cambridge MA: October Books, MIT Press.

Liberman, Alvin M. & Douglas H. Whalen (2000). On the relation of speech to language. *Trends in Cognitive Sciences* 4, 187–196.

Lieberman, Philip (1998). *Eve spoke. Human language and human evolution*. London & Basingstoke: Picador.

Lloyd, Dan (2004). *Radiant cool. A novel theory of consciousness*. Cambridge, MA: MIT Press.

Martinet, André (1949). La double articulation linguistique. *Travaux du Cercle Linguistique de Copenhague*, 5, 30–37.

Martinet, André (1957). Arbitraire linguistique et double articulation. *Cahiers Ferdinand de Saussure* 15, 105–116. Geneva: Librairie E. Droz.

Meier, Richard P., Kearsy Cormier, & David Quinto-Pozos (Eds.) (2002). *Modality and structure in signed and spoken languages*. Cambridge: Cambridge University Press.

Morwood, M.J., R.P. Soejono, R.G. Roberts, T. Sutikna, C.S.M. Turney, K.E. Westaway, W.J. Rink, J.-X. Zhao, G.D. van den Bergh, Rokus Awe Due, D.R. Hobbs, M.W. Moore, M.I. Bird, & L.K. Fifield (2004). Archaeology and age of a new hominin from Flores in eastern Indonesia. *Nature*, 431, 1087–1091.

Rizzolatti, Giacomo, Luciano Fadiga, Vittorio Gallese, & Leonardo Fogassi (1996). Premotor cortex and the recognition of motor actions. *Cognitive Brain Research*, 3, 131–141.

Saussure, Ferdinand de (1916). *Cours de linguistique générale* (ed. de Mauro). Paris: Payot 1995.

Senghas, Ann, Sotaro Kita, & Aslı Özyürek (2004). Children creating core properties of language: Evidence from an emerging Sign Language in Nicaragua. *Science* 305, 1779–1782.

Stokoe, William C., Dorothy C. Casterline, & Carl G. Croneberg (1965). *A dictionary of American Sign Language on linguistic principles*. Silver Spring, Md.: Linstok.

Tallerman, Maggie (2007). Did our ancestors speak a holistic protolanguage? *Lingua*, 117, 579–604.

Tattersall, Ian (2001). *Species diversity in human evolution*. Paper prepared for the Berlin Summer Academy on Human Origins at the Berlin-Brandenburgischen Akademie der Wissenschaften. August 20–24, 2001.

Tomasello, Michael (1999). *The cultural origins of human cognition*. Cambridge MA and London: Harvard University Press.

Trabant, Jürgen (1996). Thunder, girls, and sheep, and other origins of language. In Jürgen Trabant (Ed.), *Origins of language* (Workshop Series 2, pp. 39–67). Budapest: Collegium Budapest.

Wray, Alison (2000). Holistic utterances in protolanguage: The link from primates to humans. In Chris Knight, Michael Studdert Kennedy, & James Hurford (Eds.), *The evolutionary emergence of language: Social function and the origins of linguistic form*. Cambridge: Cambridge University Press.

Index

In the series *Benjamins Current Topics (BCT)* the following titles have been published thus far or are scheduled for publication:

12 **DROR, Itiel E. (ed.):** Cognitive Technologies and the Pragmatics of Cognition. 2007. xii, 186 pp.

11 **PAYNE, Thomas E. and David J. WEBER (eds.):** Perspectives on Grammar Writing. 2007. viii, 218 pp.

10 **LIEBAL, Katja, Cornelia MÜLLER and Simone PIKA (eds.):** Gestural Communication in Nonhuman and Human Primates. 2007. xiv, 284 pp.

9 **PÖCHHACKER, Franz and Miriam SHLESINGER (eds.):** Healthcare Interpreting. Discourse and Interaction. 2007. viii, 155 pp.

8 **TEUBERT, Wolfgang (ed.):** Text Corpora and Multilingual Lexicography. 2007. x, 162 pp.

7 **PENKE, Martina and Anette ROSENBACH (eds.):** What Counts as Evidence in Linguistics. The case of innateness. 2007. x, 297 pp.

6 **BAMBERG, Michael (ed.):** Narrative – State of the Art. 2007. vi, 271 pp.

5 **ANTHONISSEN, Christine and Jan BLOMMAERT (eds.):** Discourse and Human Rights Violations. 2007. x, 142 pp.

4 **HAUF, Petra and Friedrich FÖRSTERLING (eds.):** Making Minds. The shaping of human minds through social context. 2007. ix, 275 pp.

3 **CHOULIARAKI, Lilie (ed.):** The Soft Power of War. 2007. x, 148 pp.

2 **IBEKWE-SANJUAN, Fidelia, Anne CONDAMINES and M. Teresa CABRÉ CASTELLVÍ (eds.):** Application-Driven Terminology Engineering. 2007. vii, 203 pp.

1 **NEVALAINEN, Terttu and Sanna-Kaisa TANSKANEN (eds.):** Letter Writing. 2007. viii, 160 pp.